John A. Tarbell

Homoeopathy Simplified

domestic practice made easy - containing explicit directions for the treatment of

disease, the management of accidents, and the preservation of health

John A. Tarbell

Homoeopathy Simplified
domestic practice made easy - containing explicit directions for the treatment of disease, the management of accidents, and the preservation of health

ISBN/EAN: 9783337390655

Printed in Europe, USA, Canada, Australia, Japan

Cover: Foto ©berggeist007 / pixelio.de

More available books at **www.hansebooks.com**

HOMŒOPATHY SIMPLIFIED;

OR,

DOMESTIC PRACTICE MADE EASY.

CONTAINING EXPLICIT DIRECTIONS FOR THE

TREATMENT OF DISEASE, THE MANAGEMENT OF ACCIDENTS, AND THE PRESERVATION OF HEALTH.

BY

JOHN A. TARBELL, A. M., M. D.

Revised Edition.

BOSTON:

OTIS CLAPP & SON.

1890.

TO

SAMUEL GREGG, M. D., M. M. S. S.

THE FIRST PHYSICIAN OF THE "OLD SCHOOL"

WHO ADVOCATED

THE CAUSE OF HOMŒOPATHY IN NEW ENGLAND

𝔗𝔥𝔦𝔰 𝔏𝔦𝔱𝔱𝔩𝔢 𝔚𝔬𝔯𝔨

IS RESPECTFULLY INSCRIBED

IN ACKNOWLEDGMENT OF MUCH KINDNESS RECEIVED FROM HIM

BY

THE AUTHOR.

BOSTON, MAY 1, 1863.

PREFACE.

In the prosecution of an undertaking so important as that of popularizing the Science of Homœopathy, a wide deviation from the usual course pursued by writers of family treatises has been found necessary. An arrangement that would prevent the perplexity incident to the difficult selection of the suitable remedy from a needlessly extended list of half-proved and rarely indicated medicines, has been long thought most desirable by all who have resorted, without previous medical instruction, to the homœopathic method of curing disease. Such an arrangement is here presented. Enumerated according to the degree of their usefulness, and placed in such prominence as will insure their ready selection, will be found the few remedies that, under ordinary circumstances, and for the simple forms of disease, are likely to be indicated and best adapted for domestic dispensation. By this method of arrangement, the embarrassment and dissatisfaction almost certain to result from the perusal of several pages of reliable and unreliable, important and unimportant pathogenetic symptoms, which are recorded in confusing groups, and unattended by any well-defined marks of distinction, will be avoided, as the prescriber's attention cannot but be immediately attracted to the chief remedies, such as would, in a great majority of instances, be the most suitable for trial. The symptoms,

a *

also, when recorded as determining the use of any particular medicine, will be only such as have been found to be of the most decided and accredited character.

It is not to be denied, however, that every symptom, or assemblage of symptoms, requires a specific homœopathic remedy. But when it is considered that simple forms only of disease are supposed to be subjected to family prescribing, that uncomplicated disorders ordinarily present common features, and that the range of remedies, under such conditions, is limited, the probabilities of obtaining the right medicine, and consequently the chances of curing, are obviously very much greater than when attention is directed to a distractingly large number and variety of remedies, calculated for every imaginable contingency, with differences in symptoms so delicate as to transcend all ordinary powers of discrimination.

In the practice of Homœopathy, the principal difficulty experienced by the physician consists in the selection of a suitable remedy. During the progress of protracted, obstinate, and complicated disorders, the utmost skill and discretion of the well-educated medical attendant are called into exercise. That intimate knowledge of disease, which alone is obtained by a long course of attentive study and observation, is needed to carry certain prolonged or complex maladies to a successful termination, and to meet the exigencies that may arise in their progress. It is, therefore, ·entirely unwarrantable to assume that a non-professional person can be made capable, by the mere perusal of a domestic treatise, of prescribing, in every case, as judiciously as those who have devoted their lives to the study of disease. Unhappily for the cause of Homœopathy, and for the safety of individuals, such an assumption has been encouraged by all writers on family practice, with a single late exception ; and the impression has become quite prevalent that any one who possesses a case of medicines, with its accompanying book of necessarily limited instruction, is equally as well qualified as the physician to undertake the charge of every complaint whatever that may chance to occur, from the slightest indisposition to the most malignant epidemic.

It is with a view to remove this dangerous error, by confining domestic treatment within its proper limits, as well as to facilitate household practice by presenting concise descriptions of the various diseases, divested as much as possible of technicalities, and by a perfectly simple arrangement, to adapt to the comprehension of the unpractised reader that which has hitherto proved only perplexing and unavailable, that the present task has been undertaken. Should it prove effectual to the establishment of more correct views in relation to the non-professional management of the sick, the author labors will have been abundantly rewarded.

CONTENTS.

CHAPTER I.
DISEASES OF THE DIGESTIVE FUNCTION.

CHAPTER II.
DISEASES OF THE ORGANS OF RESPIRATION.

(1)

CHAPTER III.
DISEASES OF THE BRAIN AND NERVOUS SYSTEM.

CHAPTER IV.
ON FEVERS.

CHAPTER V.
CUTANEOUS DISEASES.

CHAPTER VI.
DISEASES OF THE EYE AND EAR.

CHAPTER VII.
GENERAL DISEASES.

CHAPTER VIII.
DISEASES OF INFANCY.

CHAPTER IX.
DISEASES OF WOMEN.

CHAPTER X.
ACCIDENTS.

CHAPTER XI.
HYDROPATHY, OR THE WATER CURE.

CHAPTER XII.
REMARKS ON HYGIENE.

A LIST OF REMEDIES

RECOMMENDED IN THIS WORK

1. ACONITE
2. ANTIMONIUM (TART).
3. ARNICA.
4. ARSENICUM.
5. BELLADONNA.
6. BROMINE.
7. BRYONIA.
8. CALCAREA (CARBONICA).
9. CAMPHOR.
10. CANNABIS.
11. CANTHARIDES.
12. CARBO VEGETABILIS.
13. CHAMOMILLA.
14. CINCHONA OR CHINA.
15. COCCULUS.
16. COFFEA.
17. COLOCYNTHIS.
18. CUPRUM.
19. DIGITALIS.
20. DROSERA.
21. DULCAMARA.
22. HELLEBORUS.
23. HEPAR SULPHURIS.
24. HYOSCYAMUS.
25. IGNATIA.
26. IPECACUANHA.
27. LACHESIS.
28. LYCOPODIUM.
29. MERCURIUS SOL.
30. MERCURIUS COR.
31. NUX VOMICA.
32. OPIUM.
33. PHOSPHORUS.
34. PULSATILLA.
35. RHUS TOXICODENDRON
36. SAMBUCUS.
37. SANTONINE.
38. SEPIA.
39. SILICEA.
40. SPIGELIA.
41. SPONGIA.
42. STANNUM.
43. STRAMONIUM.
44. SULPHUR.
45. SULPHURIC ACID.
46. VERATRUM.

INTRODUCTION.

In the year 1811, a medical work appeared in Leipsic, Germany, written by Dr. Samuel Hahnemann, in which the doctrine of Homœopathy was first presented to the world, with a detailed account of the circumstances that led to its discovery, and the long-continued and well-studied experiments necessary to establish its truth. The author had already obtained a high reputation throughout the continent of Europe as a man of science, and the medical philosophers of his time and country, were predisposed to consider, with respectful attention, whatever he thought worthy of promulgation. But the views advanced in the work above mentioned were so novel and extraordinary, so completely at variance with the prevailing impressions, and irreconcilable with the popular belief, that they occasioned less credence than astonishment. Yet when distinguished physicians began seriously to examine the overwhelming proofs in support of the theory, and after practical investigation in the sick room, publicly to declare, one by one, their unqualified adherence to the homœopathic doctrine, a degree of interest in its favor grew up in Germany and France never before equalled in the history of medicine. It was then remembered that intimations of the probable existence of such a law of cure had been given

1 *

by Stahl, Bertholon, Stoerck, Thomas Erasmus, Haller, and other eminent medical authorities, although never before practically demonstrated.

About twenty years previous to the appearance of this, his first work on Homœopathy, Hahnemann had conducted certain experiments upon himself, while in health, with the view of ascertaining the true action of cinchona, (Peruvian bark,) on the system, being induced to this trial by the perplexing and contradictory statements published respecting the curative properties of this drug. He took several doses of a strong decoction of the bark in the morning of many consecutive days, and uniformly experienced towards evening febrile symptoms, analogous in every particular to those of an intermittent fever. Astonished at the resemblance of symptoms thus produced to those which cinchona was known to remove, he was led to suppose that the cure of disease consisted in this similarity of action, depended on the principle "similia similibus curantur," or the application of remedies for the cure of symptoms similar to those which the same medicines produced on persons in health. For the purpose of verifying this conjecture, he continued to experiment with other medicinal agents, long regarded as successful remedies in certain forms of disease. Ulcers in the throat, inflammatory glandular swellings, and other effects were produced by the continued use of *mercury;* cutaneous eruptions resulted from large doses of *sulphur;* while complaints of a similar character are known to be controlled by the judicious administration of those minerals. The three medicines referred to had acquired the reputation of "specifics." By what mode of action such a character had been established; under what law other than that of "*similia similibus curantur,*" (like is cured by like,) this specific operation can be enrolled, it would puzzle any pathologist to determine. After numerous trials, under all circumstances, the consequences were the same. Different medicines produced distinct classes of symptoms, undeniable in their existence, uniform in their character. By comparing these results with subsequent experiments upon the sick, it was ascertained that convalescence became rapid and complete whenever remedies were employed, the effects of which upon the healthy corresponded with the symptoms observed in disease; and the closer this correspondence, the more immediate and decided was the recovery. Day after day, year after year, Hahnemann, with a few medical associates, pursued his investigations with uncommon perseverance and devotion, anxiously watching the development of a great truth, seizing upon every circum-stance favoring its elucidation, and sternly rejecting whatever would not bear thorough and searching criticism.

By a series of trials, unparalleled in rigid scrutiny, the founder of Homœopathy claimed to have established, among others, the following important propositions : —

1st. That a group of morbid symptoms, from whatever cause arising, will be removed by that medicine which is capable of producing a similar class of symptoms.

2d. That the curative property of medicine becomes, by the process of trituration or succession, greatly augmented; and that extremely minute doses are, in consequence, sufficient to counteract any disordered condition.

3d. That the effect of medicinal substances is destroyed by combination, and their efficiency advantageously exercised only when administered singly.

Explanatory of the first proposition, it was remarked that remedies produced a train of symptoms which were manifestations of a medicinal disease similar to the natural one, and as the existence in the human system, at the same time of two diseases resembling each other was impossible, the morbid must yield to the medicinal symptoms. The latter, dependent on the continued application of the remedy, will disappear when that remedy is discontinued, and the natural condition of the affected parts will consequently be restored. Or, again, the symptoms indicative of disease are but manifestations of the recuperative efforts of nature; and, as the medicinal agent operates in the same direction, it will aid this restorative process, and thus facilitate recovery.

Respecting the second proposition, that the potency of medicines is greatly increased by a peculiar mode of preparation, it is asserted that, in the crude state of a drug, its virtues, to a considerable extent, remain latent, and are only to be brought into full activity by a thorough breaking down of the enveloping material, and a complete separation of the molecules of matter. Only by the rapid motion of liquids and the friction of solids, is the cohesion of particles destroyed. These two processes develop a power, — as in the case of electricity, — which, without their agency, would have no existence, or, rather, would have existed in an inactive, dormant state. Whether the curative influence of medicine be disengaged by the minute division of particles, so that their mobility is increased, and placed more in affinity with the animal fibre on which they are to operate, or whether a new power is developed, like electricity, by friction, and transmitted, by successive infection, to inert substances with which it may be connected, diffusing itself to an almost unlimited extent while preserving all its primitive qualities, are questions, in the determination of which no point of practical importance is involved. Analogous facts of the disengagement of inherent force by friction, and of the immense influence upon corporeal substance of impalpable, imponderable agents, are well known and universally acknowledged.

In relation to the third proposition concerning the administration of one medicine at a time, Hahnemann's experiments proved that every medicine produced effects peculiar to itself, — gave rise to a group of distinct symptoms, — and that an assemblage of similar symptoms, when manifesting the presence of a natural disease, could be entirely removed by a single remedy. If one uncombined remedy proved sufficient to accomplish a cure, no resort need be had to a medicinal compound. Even when a complication of symptoms seemed to demand different remedies, there was no proof that a union of remedies would act in the compound as they do singly; but, on the contrary, every reason to believe that the various ingredients of which an allopathic prescription was composed, generally modify, if not completely neutralize each other, producing an effect altogether different from the one desired, and ultimately bringing about an intricacy of drug symptoms that but few physicians could comprehend, and fewer remove.

During the earlier experiments of Hahnemann and his associates in the sick room, ordinary doses, in the form of tinctures, were administered; but these doses being found too energetic, producing a temporary aggravation of symptoms, a reduction was gradually made, by the dilution of liquids and the trituration of solids, until that degree of attenuation was reached which was best calculated to remove disease safely and promptly. The dispensation of these minute quantities of medicine excited the principal opposition to the practice of Homœopathy. It was contended that such a practice was inefficient and unreasonable; that no effect could be produced by agents of such exceeding minuteness.

It was urged, in reply, that analogous results were furnished, every where, in abundance; that the smallest amount of the poison of hydrophobia, of small pox, &c., absorbed by the human system, will act with overpowering potency; that infection of all kinds, the imponderable influences of magnetism, caloric, electricity, and emanations from various sources, are in continual and violent action on animal life throughout creation. They are known, felt, and acknowledged by all, even though imperceptibly small as to material, to possess irresistible and positively destructive power. Yet the assertion of the existence of a similar property in medicinal agents plainly recognizable by chemical tests, while infinitesimal in quantity, excited the utmost surprise and incredulity, notwithstanding such agents operated directly on the delicate living texture, rendered by disease extremely susceptible to foreign impression.

But while the advancement of a reasonable theory failed to produce general conviction, the striking results of the practice, from the first trial, could not be overlooked. And whatever hypothesis might be propounded, whether admitted or rejected, the progress of Homœopathy was to depend upon facts alone. To the astonishing success of its treatment, in the most serious forms of disease, was due its establishment, and to continued success, its progress and prosperity. To what source but this can be fairly attributed its present steadfast position in every intelligent community on the face of the globe? We are aware of the numerous explanations offered by those physicians who cannot spare time for a practical investigation of its claims, — the only possible method of proving its truth or its falsity, — and they all utterly fail to account for the past and existing prosperity of the homœopathic system of practice. It has been assailed at every point, and by opponents of every name and character, and has been finally left to "die of itself." Yet it exhibits constantly augmenting activity, and at the present period of time affords incontestable evidence of possessing elements that are indestructible.

In therapeutics, or that branch of medical science which consists of the dispensation and operation of remedial measures, — to which all the other branches are subservient, and for which alone they are to be studied, — lies the grand distinction between Homœopathy and all opposing doctrines. Neither medical botany, chemistry, surgery, nor any of the means in present use to discover the character or cause of disease, are disregarded by the practitioners of Homœopathy. Its specialty appertains exclusively to the mode of administration and preparation of medicine; while the fixed and unchangeable principle on which this practice is founded, constitutes the superiority for which we contend. The history of the so-called "regular treatment" has been one of perpetual change. The remedies that were administered with confidence at one period, were altogether discarded and forgotten in another; the practice of twenty years since is not at all the practice now; the medicines then in general use were entirely unknown twenty years previous. The medical records of the past, from the time of Hippocrates to the present, are crowded with erroneous theories, ephemeral views, dangerous experiments, resulting in nothing substantial or reliable, and leaving for future generations no safe or uniform system of therapeutics. Instead of an accumulation of facts, as in all other arts and sciences, there comes down to us a cumbersome collection of distracting fictions. On the contrary, the homœopathic practice of medicine, having for its guide and basis an immutable law, grows healthily by

the aid of positive provings, that, once received, are never
rejected.   And, in the progress of time, as its advantages
are appreciated more and more widely, and the blind preju-
dice, which now withholds the privileges of civic sanction,
dies out, as it assuredly must, Homœopathy will stand acknowledged by all nations as the only true foundation of
medical science.

HOMŒOPATHY SIMPLIFIED.

CAUSES OF DISEASE.

NUMEROUS are the influences constantly in operatior within and around us tending to disturb that harmony of vital action which constitutes health. Some of these influences are inherent, capable, in many instances and to a certain extent, of modification, but seldom of removal; while others, and by far the greater number, are the result of ignorance or imprudence. Such disturbing causes are predisposing and exciting. By the former are to be understood all such as have induced a susceptibility to any form of disease, whether in action previous or subsequent to birth; by the latter, any influence which immediately impresses systems rendered thus susceptible. The distinction may be better understood by the following illustration: Several individuals, we will suppose, are subjected, at one moment, to a common influence — for example, to an abrupt change of temperature, while unprotected. Some among them, predisposed to pulmonary disease, may be attacked with cough, ending in consumption, or by pleurisy, while others will trace directly to this

2

exposure rheumatic or similar affections. The atmospheric change was the *exciting* cause; the peculiar character and seat of the complaint induced depended on the *predisposing* cause.

It is difficult to determine in all cases how far morbid influences act, either in predisposing to or exciting disease. Many causes, however, are too obvious to escape notice, the most important of which will be mentioned.

Heat and cold directly affect health, and are, especially in our climate, prolific sources of its disturbance. The most common disorders of New England, those of the throat and lungs, prevail during the winter and spring, the air passages being irritated, through the agency of cold, either by inhalation or sympathetic communication with the external surface of the body; while abdominal complaints are most prevalent during the summer and autumn, heat being found to affect the lining membrane of the stomach and intestines, through the same power of sympathy.

Sudden variations of temperature, long-continued rains, heavy dews, and certain electrical conditions of the atmosphere, are exciting causes of disease, particularly of that class termed rheumatic, the nature of which is so obscure. Vitiation of the air from vegetable decomposition produces febrile disorders of a malignant type. The quality of the blood is impaired, and scrofulous diseases, among others, result from long confinement in unventilated rooms, where the air is unfit for respiration. This is especially the case i

small, tightly-closed sleeping apartments, where there is not a sufficient supply of oxygen introduced to replace that which is consumed.

Another fruitful cause of disease lies in the use of food that is innutritious, as also in hurried mastication, irregularity and immoderation in eating. To these may be added the lavish use of condiments, which, though temporarily promoting activity of digestion, is followed, as in the case of all stimulants, by diminished energy. Dyspepsia, with its long train of evils, is chiefly confined to those who are careless in the choice, and intemperate in the use, of food.

Insufficient bodily exercise leads to disease. Congestion of the lungs and abdomen, disorders of digestion, nervous headaches, and other derangements, may be traced to this cause. The sedentary habits of students, and the inactive occupations of various artisans, induce forms of ill health unknown to those engaged in out-door pursuits that demand much physical exertion.

The swallowing of poisonous substances, or their introduction into the system by absorption or. inhalation, may be mentioned as the immediate or remote occasion of inflammation of the stomach and lungs. The atmospheric poison termed malaria, originating from decomposed vegetable matter, through the combined action of heat and moisture, gives rise to fevers of a peculiar kind (intermittent), which prevail extensively in the western and southern portions of our country. When, from modifying circumstances, this malaria does not exist in sufficient force to induce the purely intermittent fever, paroxysmal headaches and nervous difficulties, or general indisposition, result from exposure to it.

Obstructed perspiration is a very frequent cause of disease. Exposure to currents of cold or damp air, when fatigued or heated, brings on inflammation, which assumes various febrile forms. Neglect in removing, by daily ablution, the impurities accumulated on the skin, that obstruct free perspiration, more or less directly conduce to the same results.

Many physical disorders take their rise from mental disturbances. Protracted intellectual exercise, frequent unsubdued fits of passion, long-continued anxiety, undermine the health, and are the remote or immediate cause of the most severe bodily affections. Diseases of the heart and liver, convulsions, hysteria, and consumption, often have a mental origin. Instant death by apoplexy has been induced by anger. Prolonged fainting fits, terminating fatally, have been caused by fear. Jaundice, and even erysipelas, have been brought on by rage and indignation. The most obstinate forms of indigestion, and organic diseases of the stomach, have been produced by mental anxiety and strong emotion.

Allusion has been made, in general terms, to poisons, as among the most important causes of disease. In connection with that class may be mentioned one as well established as any other, and much more common, viz., medicinal drugging. The very measures adopted by allopathy to cure disease have been the most efficient in its production. There exists no doubt whatever that many disorders of the most obstinate and distressing description owe their origin solely to the action of medicine administered in poisonous doses. Years may elapse after the absorption of the poison before its influence ceases. The noxious effects of arsenic, mercury, and opium, in their several forms, together with other mineral and vegetable preparations in use as remedies, are obvious, and not unfrequently last for life. However studiously this prolific source of disease has been, and is, overlooked by many who choose not to recognize it, its reality is none the less absolute and demonstrable.

Among the strictly predisposing causes of disease may be reckoned hereditary tendencies. It is very evi-. dent that pulmonary and scrofulous complaints are, in many instances, transmitted from parents to children. There is no doubt that cutaneous diseases, cancer, gout, insanity, and inclination to hemorrhages, to deafness, blindness, &c., have been inherited. Such inheritances demand the persevering, uninterrupted use of those means of prevention best adapted to individual cases, and to which reference will be made in another place.

The cause of many very serious diseases is still involved in obscurity, and seems, as in the case of Asiatic cholera, placed beyond the limits of discovery. Many affections of great severity are confined to certain localities (endemics), and are attributable to circumstances peculiar to the place in which they appear: thus, the "goitre," a tumor on the fore part of the neck, is often

2 *

met with in Derbyshire, England, and in certain can-
tons of Switzerland, but is seldom seen elsewhere;
fever and ague prevails in some of the United States,
and not in others; a virulent kind of ophthalmia in
Egypt; with many other affections in different localities, the causes being sometimes traceable, but oftener
undiscovered. So with "epidemic" diseases, that appear in certain seasons, generally without a known
cause, and not unfrequently occasioning great mortality. They are distinguished from "endemics" by not
being limited to particular situations, but extending
from place to place, and often being communicated by
one to another. The yellow fever, cholera, plague,
many malignant dysenteries and fevers that are classed
among the epidemics, have, in respect to their origin,
baffled investigation; and though numerous causes
have been assigned, none have in all points been regarded as satisfactory. Many endemic and epidemic
disorders, however, have recognized sources, among
which may be enumerated unwholesome food, of un-
matured grain and vegetables, tainted meat, unripe
fruit, and diseased fish, certain low, damp situations,
districts exposed to great heat, alternately covered by
water and left exposed, stagnant water, impure air, the
crowding of many individuals in a small space, malaria.

The last and most important cause to be mentioned
is contagion, by which disease, originating, it may be,
spontaneously, is communicated from one individual to
another. Infection, a term frequently used as synonymous with *contagion*, is more applicable to the epidemic

influence alluded to in the preceding section under the name of " malaria," while contagion, in the strict etymological sense of the term, implies contact, and should be used only to express the transmission by touch of disease from one person to another. Still, the terms are capable of modification, and are not, in every view, altogether explicit. Infection may be considered as acting through the medium of the atmosphere, whether emanating from animal or vegetable substance, and produces, in such as are exposed to its influence, intermittent and remittent bilious and yellow fevers, dysentery, &c. The exclusively contagious diseases are few in number, being limited to cutaneous affections and others that are seldom met with in temperate regions. The small pox, chicken pox, scarlet fever, measles, and erysipelas, are communicated through the medium of the air, as well as by contact.

The disease-producing agents above named are known only by their effects, and under all circumstances the identical character of each is preserved. They are termed " aura," emanations, exhalations, &c. They are invisible, impalpable, imponderable. Whatever of material, if any, the noxious agent possesses, — whatever be its nature or its name, — it certainly enters the human system in strictly *infinitesimal potencies,* and thus, in a state of extreme dilution, manifests its presence by decided symptoms, and if not combated with similarly subtle agents, acts with an energy that often proves destructive to life.

SYMPTOMS OF DISEASE.

EVIDENCE of the character and extent of those disturbances and alterations that constitute disease is
derived from two sources.— from information received
through the patient's description of his own sensations,
and that which is obtained by observation of external
manifestations. The former are called " rational " and
the latter " physical " symptoms. In diseases of very
young children, and of those who are deprived of
speech, or such as can not speak in language known
to us, the physical symptoms are the only means of
information at our command.

Of the rational symptoms, pain is the most frequent
and important. It is present, in its various degrees, in
nearly all disordered states of the body. Its character
and location often go far in disclosing the true nature
and extent of the disease existing, although it is sometimes a symptom, independently regarded, on which not
much reliance can be placed. The terms made use of
by the sufferer to express pain are not unfrequently
deceptive, being either extremely indefinite or highly
exaggerated. Much allowance must be made for temperaments, in judging of the actual amount of pain.
In addition to this symptom there are others that

properly come under the head of rational signs, as nausea, giddiness, want of appetite, thirst, derangement of sight, hearing, and feeling, sensations of oppression, of burning, itching, sinking, &c. Every uncomfortable bodily sensation indicates disorder of some kind, and to some extent, since absolute health must be considered as a condition wholly exempt from every species of disturbance.

Without more special reference in this place to the sensations above named, such external signs will now be mentioned as are most prominent and valuable in designating different morbid conditions. The appearance of the face, the color and temperature of the skin, the state of the tongue and eyes, the attitude of the body, the pulse, the respiration and voice, with other physical symptoms, are of value in representing, to a certain extent, by their deviations from the normal condition, the commencement, progress, and probable termination of the disease. It will not be necessary at present to designate every particular sign, that, under all circumstances, is to be regarded as indicative of disease, but allusion will be made to such as are the most noticeable and important.

An unusual redness of the face attends inflammatory fever, inflammation of the brain, apoplexy, and certain eruptive diseases. A distinctly circular redness of the cheeks is indicative of the hectic fever that attends pulmonary consumption. Paleness of the face is present during low fevers, after loss of blood, dropsy, scrofula, &c. Sudden and frequent changes from paleness to redness, and the reverse, attend acute dropsy of the brain. Yellowness of the face denotes a disturbance of the liver, or an obstruction of the ducts leading thereto. It is seen, though less distinctly, in cancerous and certain other affections of the stomach. A livid color of the face occurs in diseases of the heart, sometimes in inflammation of the lungs, in apoplexy, and in the last stages of cholera.

The temperature of the skin is higher than natural in all acute diseases accompanied by fever. The heat is greatest in scarlatina, and in some local inflammations. Hectic fever is attended with heat in the palms of the hands and the soles of the feet. Heat on circumscribed portions of the surface indicates an inflammatory or highly irritable state of the parts adjacent or immediately contiguous. A diminution of the natural temperature attends many chronic diseases, and is indicative of debility, except, as is at times the case, when the result of powerful mental emotion. Chills attend the first stage of some fevers, and especially of the intermittent. If redness exists with swelling, and accompanied with heat or pain, a superficial inflammation is present. If the redness is dark, circumscribed, and small blisters are seen over the swollen surface, erysipelas is denoted. Should the swelling be white, and receive from pressure a marked indentation, there is a collection of water under the skin. Should it be white, and receive no such indentation, but be elastic, and give out, on pressure, a kind of crackling sound, there is air beneath the skin.

The state of the tongue is of much importance in its association with other indications of disorder. The unnatural secretion of its mucous membrane, or, as it is termed, the "coating," presents many and various appearances. A thin, white coating is the one most frequently observed, and occurs in the diseases of the stomach and lungs, in rheumatism, and inflammations. It may be said that inflammatory affections are indi· cated by it, while a much thicker, white, lard-like coating appears in diseases of a typhoid character, in malignant fevers, in severe forms of scarlet fever and small pox, and in most diseases that are attended with great debility. A yellowish coating is generally observed in disorders of digestion, in diseases of the liver and bowels, and in diarrhœa ; sometimes the coating is green under similar circumstances. A brown coating is perceived in chronic disorders of the bowels, in hemorrhoids, in chronic jaundice, in the advanced stages of typhoid and putrid fevers, or when the original white coating has been darkened by the action of inflammation. When the tongue appears nearly black, it indicates an extremely vitiated condition of the blood, and this color must be regarded quite unfavorable as a symptom, although sometimes manifest where no great danger exists. If the tongue, previously coated white, becomes red, it is symptomatic of violent inflammation of the stomach or bowels, sometimes of the lungs. When the coating, of whatever character it may have been, passes off, leaving the tongue of a natural color, clean and moist, it is one of

the most certain evidences of convalescence. There
is a difference to be observed in the cleaning of the
tongue. If the coating should appear to be passing
off from the edges towards the centre of the tongue,
the indication is favorable for a speedy convalescence.
If from the centre towards the edges, the former becoming first clean, convalescence will be comparatively
slow.

The appearance of the eyes with regard to color,
position, motion, and expression, deserve notice. They
are discolored in many diseases, either local or general;
are red in ophthalmia, in measles, and disorders of the
brain; yellow in jaundice, brown in low types of fevers,
and present every shade of color, at times, during the
progress of exclusively ophthalmic affections. The
pupil of the eye is dilated when there exists pressure
on the brain, and when also there is irritation from
sympathy with the stomach. It is contracted in inflammation of the membranes covering the brain, and
also in inflammation of the eye itself. There is to be
seen a rapid movement of the eyes in certain spasmodic affections, and a total immobility often in dropsy
of the brain. They are sometimes turned inward in
the latter case. Distortion of the eyes, when occurring at the commencement of any acute disease, is an
indication that the illness will be long-continued and
severe. An expression of wildness in the eye is seen
in disorders of the brain; one of anxiety characterizes
affections of the heart; one of despair attends serious
stomach and abdominal inflammations, and is generally

observed in cholera. There are many derangements of vision, such as the view of half an object, the distortion, inversion, and multiplication of images ; which derangements, when other than evanescent and temporary, are owing chiefly to structural diseases of the eye, but may accompany various affections of the brain, from the disorder of slight delirium dependent on nervous excitement to acute cerebral inflammation.

The attitude, being that voluntary or involuntary position which the sick person assumes, is to be regarded as another not unimportant condition. An erect posture, from inability or unwillingness to lie down, is adopted in asthmatic sufferings, in diseases of the heart and lungs, and in dropsy of the chest. Lying flat upon the back, with a tendency to settle towards the foot of the bed, is indicative of great debility, as in the typhus and last stages of other fevers Drawing up of the lower limbs towards the abdomen is a position taken, especially by children, in inflammation of the bowels. A perfectly quiet posture, retained for some time, if combined with complete unconsciousness, is indicative of pressure on the brain. An entire relaxation of the muscles is observed in fainting and in apoplexy. A rigid state of the muscles prevails in certain spasmodic affections, in catalepsy, and in that disorder involving the whole body — " tetanus " — which, when confined to the muscles of the mouth, is known as " lockjaw." When general rigidity is observed in the advanced stage of low fevers, it is a symptom suggestive of great danger, especially if attended with spasmodic tremulousness of the tendons at the wrist. A great degree of restlessness, indicated by tossing the arms about constantly, is, when observed to occur after exhausting illness, to be considered as of serious import.

The pulse, that very important indicator of the state within, the sign to which the physician first resorts for guidance, claims special attention. Situated superficially, and at a point the most accessible, the radial artery, which is the "pulse," represents the number of the heart's contractions. Its average healthy pulsations are 75 per minute in adults, and in children from 80 to 120, the latter number being that of the infant's pulse. In febrile and other affections of infancy, the pulse indicating inflammation would be above 120; the feverish pulse of the adult above 75. With many individuals, the pulse, from some peculiarity in no degree dependent on health, is habitually slower than 75, sometimes being 60, in rare cases as low as 40. When, however, such slowness of pulse is not constitutional, it is indicative of disturbance, sometimes simply "functional," — a term applied to deranged action that does not involve the loss or alteration of substance, — sometimes it is a sign of pressure on the brain. In old age a slowness of the pulse is natural. It is important to remember that the pulse is faster after exertion, and faster in the erect than in the sitting posture; in the sitting than in the recumbent posture. A pulse that continues constantly above 120 in adults indicates serious disease, and when as high as 150, especially if feeble, is a sign of great danger. It does, at times, amount to 100, and even to 200 pulsations; in the latter case it is with difficulty counted, not only by reason of its rapidity, but of its feebleness, the two conditions being inseparable. When the pulse beats strong and feels hard under the finger, inflammation of some kind exists. On the contrary, when it is feeble and easily compressed, there is an opposite state — a deficient action somewhere. In both cases there may be frequency of pulsation, so that the quick pulse is not, in itself, determinate. A full pulse is one in which the artery seems large in circumference, and is usually connected with the strong, hard pulse, indicative of inflammation. A small pulse, on the contrary, is narrow and thread-like; if at the same time hard, it is termed " wiry." An irregular pulse is one in which the beatings occur at unequal intervals, and are of unequal force, some being more feeble and rapid than others; occasionally there is an entire absence of one or two pulsations. This latter intermittent character is not of a decisive nature, as it may take place when there is disease of the heart, of the head, of the stomach, or in consequence of simple debility; when there is serious organic difficulty, or when there is no disease at all. In cases of extreme prostration, the intermittent pulse is an unfavorable symptom. A regular pulse is an indication generally favorable, though instances have been known of irregularity occurring during health, while in the same persons, during illness, the pulse is perfectly regular.

It requires much experience, with no little discrimination, to acquire all the information that the pulse is capable of affording ; and it would not be of much practical use here to refer to the many varieties of pulse supposed to be characteristic of certain morbid states, as there are many very fanciful and quite unnecessary distinctions.

The state of the respiration is another condition of consequence in the formation of a correct " diagnosis," —a technical term signifying the opinion obtained of the nature of a disease by an examination of symptoms. The act of inspiring and expiring air occurs, during health, about fifteen times in a minute. This action of the lungs is increased when any obstruction exists to the free entrance of air, whether in the organ itself, or in the passage leading thereto. A portion of the lungs may not be capable of free expansion on account of inflammation or its consequences, — tubercular and other morbid products, — or there may be too great an amount present in the lungs to be acted upon. In inflammation of the lungs, or of the lining membrane, called " pleura," respiration is greatly accelerated, sometimes amounting to sixty in a minute. An unnaturally slow respiration is observed in cases where the heart's action is diminished, in a state of unconsciousness from poisoning and other causes, and in the hypochondriac. When respiration becomes laborious, there is no pause between the acts of inspiring and expiring air, as is the case in health, and all the muscles that can be made to assist respiration, as in croup, are called into action. The muscles of the chest are in motion, and these only, when there are serious affections of the abdomen; and when, from disease of the lungs, both sides of the chest are incapable of respiration, this action is performed wholly by the abdominal muscles. Irregularity of breathing is connected with diseases of the brain and spasmodic affections. Respiration, like the heart's pulsations, sometimes intermits, which is, in brain diseases, an unfavorable symptom. Stertorous or snoring respiration attends apoplexy. Many of the distinctions made with reference to respiration, as in the case of pulsation above alluded to, are of no considerable practical importance.

There are signs other than those enumerated, to which it will be proper here to allude, although, in order that such means of gaining information be fully available, much previous study is necessary. We refer to percussion and auscultation. A definition of the terms is all that need be now given. Percussion denotes the striking on the chest with the hand to ascertain by the sound elicited the condition of the part struck. The healthy lungs, being filled with air through their whole extent, will give out, when the chest is percussed, a hollow sound, like that produced on striking an empty barrel. If the lungs, in certain parts, are rendered by disease impermeable to air, the parts so diseased will not sound hollow, but yield a dull sound, like a solid substance, when struck. In proportion as air or solid substance predominates, the

3 *

sound produced on percussion will be more or less hollow; and by this sign the condition of the lungs is, to a certain extent, ascertained. In addition to this method of obtaining information with regard to morbid changes occurring internally, there is another derived from the sound that the air produces on entering the lungs, as well as from the vibration of the voice, &c. Such sounds are best heard by the direct application of the ear to the surface of the chest, although an instrument called a "stethoscope"— being a hollow cylinder made of wood, — has been in frequent use, and is in some cases to be preferred. This mode of examination — auscultation — is applicable to the investigation of diseased states of the heart as well as of the lungs. The natural sounds of organs in motion being known, all the deviations caused by disease are recognized by the educated ear.

TREATMENT OF DISEASE.

THE history of medicine in relation to therapeutics, or the treatment of disease, from the time of Hippocrates to the present century, presents a lamentable record of contradictory practices, born of false and unphilosophical speculations, that have been " promulgated by men who had a reputation to establish," and adopted by routine imitators, who blindly follow, regardless of consequences, wherever theory leads. The doctrines of the " chemist," the " mechanist," the " solidist," the " humoralist," the " vitalist," the " mathematician," the " metaphysician," those of Cullen, of Brown, of Broussais, and others, have sprung up, each influencing the entire medical practice of their time, soon to fall into utter discredit, while, after all these life-sacrificing revolutions, other arts, and especially the collateral sciences, meantime constantly progressing, the treatment of the sick by the " soi-disant " *regular* physician of the present day is not more successful than that of the " Father of Physic," who practised medicine more than four hundred years before the Christian era. In view of such a past, and of the present equally chaotic condition of the " old school," bred in and living by the grossest empiricism, — in view of the well-known fact that " retiring " practitioners are every where loudly expressing dissatisfaction with their own experience, — it is not to be wondered at that a very general scepticism now prevails in relation to the efficacy of medicine. It must be evident that the " art of healing " has not kept pace with other advancing sciences, because it has not been constructed on the unchangeable basis of a natural law ; and the amount of material that the devoted labors of learned and most benevolent men have accumulated is valueless, for want of a solid foundation on which to rest, as well as on account of the preponderance of fictions over facts. To this point the following observations of Dr. Rush are expressly directed : —

"It seems to be one of the rules of faith in our **art,** that every truth must be helped into belief by some persuasive fiction of the school. And I here owe it to the general reader to confess, that as far as I know, the medical profession can scarcely produce a single volume, in its practical department, from the works of Hippocrates down to the last made text book, which, by the requisitions of an exact philosophy, will not be found to contain nearly as much fiction as truth. This may seem so severe a charge, against both the pride and logic of our art, that I crave a moment of digression upon it.

"There are tests for all things. Now, a dangerous epidemic always shows the difference between the strong and the weak, the candid and the crafty, among physicians. It is equally true that the same occasion displays, even to the common observer, the real condition of the art, whether its precepts are exact or indefinite, and its practice consistent or contradictory. Upon these points, and bearing in mind that we have now, in medicine, the recorded science and practice of more than two thousand years, let the reader refer to the proceedings of the so-called "Asiatic Cholera," and he will see their history every where exhibiting an extraordinary picture of prefatory panic, vulgar wonder, doubt, ignorance, obtrusive vanity, plans for profit and popularity, fatal blunders, distracting contradictions, and egregious empiricism; of twenty confounding doctors called in consultation to mar the sagacious activity of one; of ten thousand books upon the sub-

ject, with still an unsatisfied call for more; of experience fairly frightened out of all its former convictions;
and of costly missions after moonshine, returning only
with clouds.

" Now, I do assert, that no art which has a suffi-
ciency of truth, and the least logical precision, can
ever wear a face so mournfully grotesque as this. In
most of the transactions of men there is something
like mutual understanding and collective agreement
on some point at least; but the history of the cholera,
summoned up from the four quarters of the earth,
presents only one tumultuous Babel of opinion, and
one unavailing farrago of practice. This even the
populace learned from the daily gazettes; and they
hooted at us accordingly. But it is equally true, that
if the inquisitive fears of the community were to bring
the real state of professional medicine to the bar of
public discussion, and thus array the vanity and inter-
ests of physicians in the contest of opinion, we should
find the folly and confusion scarcely less remarkable
on nearly all the other topics of our art.

" Whence comes all this? Not from exact observation, which assimilates our minds to one consenting
usefulness; but from fiction, which individualizes each
one of us to our own solitary conceit, or herds us
into sects for idle or mischievous contention with each
other; which leads to continual imposition on the public, inasmuch as fictions, for a time, always draw more
listeners than truth; which so generally gives to the
mediocrity of men, and sometimes even to the palpably

weak, a leading influence in our profession, and which
helps the impostures of the advertising quack, who,
being an unavoidable product of the pretending theories of the schools, may be called a physician with the
requisite amount of fictions, but without respecta-
bility."

Such is the humiliating representation afforded by
a distinguished medical philosopher of the unsettled
condition of the prevailing school of medicine. It in
no degree applies to homœopathy. A long-continued,
scrutinizing, and widely-extended experimentation has
furnished us with an array of facts that are to be relied
on by all persons in every region and circumstance.
The direct specific action of each individual medicine
made use of has been ascertained by an unprecedented
and combined industry of investigation undertaken
upon the healthy system. The pathogenesis or specific
operation of medicine on certain textures being determined, a fact of vast importance is thereby gained in
therapeutics. Without this knowledge, the physician
prescribes at random. Diseased textures are made
evident by physical and rational signs. These signs
or symptoms determine the selection of a homœopathic
remedy, on the principle of "*similia similibus curantur*," or the treatment of disease by remedies that are
capable of producing similar symptoms. The very
few specifics known for centuries were found by Hahnemann to be governed by this law of resemblance.
And the inference drawn, that artificial disease, or that
caused by medicine, specially attacks tissues affected

by the natural morbific influence, and overpowers the latter, which artificial disease in its turn is overcome by the restorative reaction induced or promoted by the medicine, is legitimately derived from the observation of facts, in accordance with the Baconian plan of induction, the only solid foundation for the establish ment of a medical truth.

The medicines administered in the practice of homœopathy are small in amount, in conformity with the general principle verified by well-conducted experiment, that the greater the affinity or relation a medicine holds to a malady, the smaller is the quantity required for its removal. Disease is attended with a great exaltation of sensibility, and a very slight impulse of a truly remedial character is favorably received, while an unnecessarily powerful impression has been found to aggravate the irritability already existing, and thereby obstruct instead of promoting restorative action. It is by gently operating in accordance with the restorative effort constantly being made by Nature, that art can *ever* minister to advantage. When otherwise engaged, it is always intrusive, oftentimes destructive. Far safer is it for the sensitive sick to be left alone with unassisted Nature, than to be subjected to the rude, misdirected violence of what is well termed a " bold practice."

ON THE ADMINISTRATION OF MEDICINE.

It will be perceived that, throughout this work, the
medicines recommended for administration are in the
form of "*globules.*" These globules are composed
sometimes of sugar and starch combined, sometimes
of "sugar of milk," a crystallized evaporation from
milk whey, this being the fittest vehicle, as it is the
purest non-medicinal substance procurable, and they
are saturated with that medicinal preparation which
has been regarded as the most generally efficient.
The globular form is selected for domestic practice as
more convenient than powder or dilution, and the
best adapted for the preservation of the medicinal
property with which it is charged.

The globules may be applied to the tongue in a dry
state, or they may be dissolved in water. In mild
cases, or in those of a chronic nature, they are generally to be used· undissolved — from three to six
constituting the usual dose. In acute diseases, with
active conditions, as pain or fever, the globules are to
be thoroughly dissolved in pure water, and a spoonful
of the solution given every two or three hours, or at
other intervals, longer or shorter, according to the
severity or mildness of the symptoms present. Seldom,

however, is it necessary to repeat the dose more frequently than once in three hours. · If improvement follows the first administration, and is progressive, a repetition of the dose will serve only to disturb the progress of the improvement, and, of course, to retard recovery.

In case other preparations of the medicine, as the liquid or powder (dilution or trituration), should be at hand, and preferred, through convenience or other cause, to the globules here recommended, they may be used as follows: Dissolve two drops of the dilution, or as much of the trituration, as will cover one fourth part of a small penknife blade, in a tumbler half filled with water, thoroughly stirring the solution, and administer doses in the manner and according to the circumstances above alluded to.

The medicine should be taken at least half an hour before, or an hour after, a meal. Urgent symptoms will not admit of delay, but when such are not present, the most favorable time for taking the medicine is on retiring to rest, as counteracting influences are less liable to be in action for several subsequent hours. In many chronic complaints, one dose daily is sufficient. It sometimes happens that existing symptoms are ag gravated after a medicine has been taken, and, in such cases, the aggravation will soon be followed by improvement. This medicinal aggravation is frequently an indication that the suitable remedy has been administered, and that it will prove ultimately beneficial. When no change, either for the better or

4

the worse, is perceptible, and no neutralizing agency is known to exist, the right medicine has not been given, and the next in the order of importance should be selected. If a violent medicinal aggravation occur (a circumstance exceedingly rare), the antidote to the remedy should be given, and afterwards the same remedy repeated in a smaller dose.

An alternation of medicines that are similar in their action has, as will be seen, been frequently recommended, and is always useful when one medicine appears insufficient for the removal of all symptoms.

The different susceptibility of individuals to medicinal impressions is to be taken into consideration. Some require larger doses than others. From two to eight globules may be given, according to circumstances. Children require less than adults; females less than males; feeble, delicate persons less than the strong and robust.

The vials containing the globules should not be exposed to heat, or to the direct rays of the sun. They should, after use, be closely corked, and immediately returned to the box.

REGULATION OF THE DIET.

Every one under homœopathic treatment should carefully abstain from all articles of food in any degree medicinal, should eat moderately of that only which is digestible, and drink water in preference to any other liquid if accustomed to it. But if unaccustomed to cold drinks, he must not change too suddenly, but may use the various preparations of cocoa, weak black tea, or coffee made from roasted farinaceous substances. Under some circumstances, milk. Water sweetened with sugar, raspberry or strawberry sirup, solutions of simple fruit jellies, barley water, rice water, gum arabic water, thin gruel, or arrowroot may be used, if desired. The drinks that should be avoided are all kinds of wine, alcoholic and malt liquors, coffee, green tea, soda water, mineral waters, and all beverages that are prepared with acids.

The articles of food allowed are, of meats, beef, mutton, poultry, wild game, fresh fish; of vegetables, potatoes, squashes, beets, turnips, tomatoes, carrots, beans and peas, excepting in case of colic or diarrhœa, when vegetables and fruits — none of which latter are usually prohibited — should be abstained from.*

All kinds of ripe (not too fresh) bread which do not contain soda or saleratus, are allowed; also puddings made of flour or bread, Indian or rye meal, oatmeal, pearl barley, rice, sago, and tapioca.

Butter that is fresh may be used sparingly, as well as cream, and the simple preparations of milk. Plain custards, soft boiled eggs, and oysters, excepting in diarrhœa or colic, are not prohibited.

It will be observed that in the above list is included a sufficient variety of aliment, not only to sustain life and strength, but to satisfy any reasonable appetite.

---

* It is of the utmost importance that in each individual case the food should be as judiciously chosen as the system may require and can bear.

Certain kinds of food in the list which follows are prohibited, not as being in themselves decidedly injurious to the system, — or, at least, not directly so, — but as tending to counteract, by the possession of indigestible or medicinal properties, the homœopathic remedies that may have been administered.

The objectionable articles are, of meats, pork, veal, salted meat and salted fish, shell fish, fish without scales, the flesh of very young animals, geese, ducks, sausages, and the liver, heart, or lungs of animals.

Every description of vegetable not included among those allowed in the preceding section, especially if pickled; all condiments, comprising the whole family of spices; all flavors from oil or essence; all acids; all distilled and fermented liquors; all confectionery; all aromatics; all kinds of pastry; and all kinds of nuts, are prohibited.

The above-named articles are excluded from the diet of those immediately under treatment. There may happen cases, however, where entire abstinence in this respect will not be altogether necessary; but deviations from the dietetic arrangement here given should not be made unless under the direction of a physician.

GENERAL DIRECTIONS.

During fevers and acute diseases generally, the diet should be as simple as possible. The entire absence of a desire for accustomed nutriment is a plain indication that it is not needed, and that the digestive organs are not in a proper condition to perform their office. Gruel, barley water, or arrowroot, and sometimes weak tea or cocoa, is the form of nutriment best suited to the enfeebled digestion of those suffering from fever. During convalescence, the return to animal food should be gradual, commencing with beef tea, mutton or chicken broth, care being exercised in relation to the quantity taken at one time.

Tooth powders containing medicinal substances should not be used. No medicines of any kind ought to be taken while under homœopathic treatment, except such as are prescribed in and accompany this work. No perfumes, as cologne, camphor, &c., should be used. The apartment of the sick should be kept quiet, and free from every thing annoying and gloomy, — should be large and well ventilated, and as light as the patient can bear comfortably; the patient's mind should, as much as possible, be kept undisturbed and tranquil. Frequent bathing with cloths wet with cold water is, in most cases, highly refreshing in febrile affections, and is seldom contraindicated when the surface of the body is dry and warm. Medicated baths are forbidden, as well as hot water baths. Bouquets of strong-smelling flowers should not be brought into a sick room, neither should growing plants be permitted to remain there, more particularly during the night.

4 *

CHAPTER 1.

DISEASES OF THE DIGESTIVE FUNCTION.

1. Teething.
2. Toothache.
3. Tic Douloureux.
4. Aphtha.
5. Mumps.
6. Inflammatory Sore Throat.
7. Ulcerated Sore Throat.
8. Dyspepsia.
9. Inflammation of the Stomach.
10. Spasms of the Stomach.
11. Vomiting of Blood.
12. Colic.
13. Heartburn.
14. Constipation.
15. Diarrhœa.
16. Dysentery.
17. Cholera.
18. Asiatic Cholera.
19. Hemorrhoids.
20. Jaundice.
21. Inflammation of the Liver.
22. Inflammation of the Spleen.
23. Inflammation of the Bowels
24. Worms.

A GENERAL view of the nature and office of the organs employed in the performance of digestion will be given preliminary to a special description of those diseases that disturb this important process; and a similar arrangement will be adopted with reference to the other classes in which it is proposed to divide the present work.

The conversion of food into nourishment commences with mastication, or the act of chewing, which is the proper preparation for its reception by the stomach

(42)

During mastication the salivary glands, as they are termed, the largest of which are situated below the ears, at the angle of the lower jaws, send through their small canals the secretion (saliva) which is intended to facilitate swallowing, as well as to prepare that which is swallowed for easier digestion.

Passing down the throat, or œsophagus, the food is introduced into the stomach, where it is subjected to the action of another secretion — the gastric juice.

This solvent, together with a 'muscular 'contraction of the stomach upon its contents, converts the food into a homogenous pulp or paste, called chyme. Then this softened material gradually passes out of the stomach into the duodenum, or first portion of the intestine, where it is again acted upon by secretions from the liver and pancreas, and changed into a fluid state denominated chyle, in which form it is absorbed, as it passes slowly on, by thousands of minute vessels, the mouths of which line the inner membrane of the alimentary canal, that is continued from the stomach. The chyle is conveyed through these small vessels into the blood ; and the latter is distributed, with its nutri tious particles, to every portion of the frame. The residue of the material, which is indigestible, and cannot be converted into nourishment, is rejected as worthless.

The entire process of digestion, comprising the three acts of mastication, chymification, and chylification, is during health accomplished in about three hours. The main part in this process is performed by the stomach, and while this organ is actively engaged, it concentrates in itself an unusual amount of vital energy; the blood is attracted to it from the other parts of the system, that it may be enabled to carry on its operations vigorously; hence chilliness of the surface is frequently experienced after a hearty meal, and also a strong tendency to sleep, especially by such as have become debilitated through fatigue or age. The operation of digestion is obstructed by substances upon which that remarkably active solvent, the gastric juice, has but little or no power, such as the seeds of plants, the rinds of certain kinds of fruit, &c. These substances often remain in the stomach for a long time, giving rise to very uncomfortable sensations; or, passing undissolved through the intestinal canal, sometimes cause general disturbance by their mere presence, and sometimes form centres for concretions which greatly endanger life. In addition to those absolutely indigestible substances which cannot be masticated, and therefore should on no account be swallowed, there are articles of food in common use that disturb the act of digestion, and are to be especially avoided by those whose digestive organs have been by any cause debilitated.

The dietary directions for the sick under homœopathic treatment have been written with reference as well to the comparative digestibility of different substances, based on Dr. Beaumont's experiments, as to their counteracting medicinal properties.

There are other organs than the stomach concerned

In the process of digestion. The bile secreted by the liver, and the secretion from the pancreas, assist in the preparation of the substances swallowed for incorporation into the system.

It would occupy too much space to pursue the subject of digestion in all its relations; to allude to the diversity of structure presented in the different classes, the general economy of their nature, their operations, and the many theories advanced in relation thereto. Sufficient, it is presumed, has been stated to prepare the reader for the better understanding of that order of diseases, a particular description of which is to be given under their respective names. And the first of this class, to which attention should be directed, is

TEETHING. (*Dentition.*)

There are three ages during which the human teeth pass through the gums — infancy, childhood, and adolescence. The process is frequently attended with considerable pain; but as serious disturbances are caused by it only during the period of infancy, when the developing system is exceedingly tender and susceptible, our remarks on the treatment of dentition will be confined to that period. The first dentition begins about six months after birth, and is generally completed before the end of the second year.

The full set consists of twenty teeth, — ten in each jaw, -- which are shed between the sixth and twelfth year, usually, and are replaced by the permanent

teeth, amounting to twenty-eight, to which are added
four others between the twentieth and thirtieth year.
In the fourth month of infancy, the rudiments of the
first teeth that appear change in the gum from a soft,
pulpy state into solid material; and it is at the time of
this change that the child appears ill, is fretful, starts
in sleep, has febrile symptoms, hot skin, thirst, restlessness, it may be drowsiness; the irritation often extending to the surface of the body, producing a rash, and
to the bowels, producing diarrhœa; sometimes to the
lungs, causing cough; sometimes to the brain, causing
convulsions.

When the teeth are on the point of protrusion,
known by the florid, distended appearance of the gum
at the base, and a paleness at the edge or upper part,
the gum should be lanced, if the symptoms are severe,
and are not mitigated by medicine. By this method
the tension is removed, and the tooth freed immediately.

First of all, in this, as in many other complaints
hereafter described, *Aconite* is to be given whenever
there exists fever, indicated by heat of the skin, thirst,
rapid or full pulse, sleeplessness, restlessness, and want
of appetite. This medicine is a specific for general
febrile action, when resulting from cold, and is to be
continued so long as the symptoms of fever are manifest. It is the principal remedy in a great number of
diseases, at their commencement, for the larger number
are attended with irritation or inflammation, the external signs of which are the symptoms above named.

The medicines adapted to treatment, with their characteristic indications, will be referred to now, and throughout this work, in the order of their importance, while others of less value, though at times possibly re quired, will be named for reference.

First after *Aconite*, the symptoms for which are named above, comes

1. CHAMOMILLA.

Feverishness, with perspiration and thirst; peevishness; convulsions; movements of the limbs; sleeplessness; diarrhœa and pain in bowels; restlessness.

2. BELLADONNA.

Flushed face and feverishness; starts during sleep, and frequent waking with anger or fear; dry cough, with hurried breathing; convulsions; irritability; pupils dilated; growing worse towards night.

3. COFFEA.

May be tried if there be much nervous excitement and wakefulness.

4. CALCAREA.

When the appearance of the teeth are long delayed, and the attendant troubles continue, or frequently recur.

Remedies for consultation, if necessary:—

NUX VOMICA; MERCURIUS; IGNATIA; OPIUM.

Administration of Medicine. Eight globules of the medicine selected are to be dissolved in eight tea-spoon-

fuls or two table-spoonfuls of pure water, and one tea-
spoonful is to be given twice a day, morning and evening, excepting in case of *Aconite* being used for febrile
symptoms, when the above dose should be repeated
every second, third, or fourth hour, until a marked
improvement is perceptible.

TOOTHACHE. (Odontalgia.)

When pain evidently results from exposure of the
nerve in a carious tooth, the surest, and sometimes the
only mode of relief is a prompt application to the skil-
ful dentist. The causes of toothache, however, are
various; sometimes rheumatic, sometimes catarrhal,
sometimes nervous, through sympathy with disturbances going on elsewhere, sometimes the deposit of
ossific matter on the sides of the tooth or its socket.
There are a great number of medicines recorded as
occasioning pain in the teeth; but since our chief
object as before stated, is to obviate the confusion
arising from a perusal of the manifold symptoms of
every variety that encumber family treatises, we shall
only refer to those medicines which have been found
most frequently curative.

1. **MERCURIUS VIVUS.**

Shooting pain passing over side of face and head;
swelling and inflammation of gums; pain increased
during the night; pain aggravated by cold food or
drink.

2. NUX VOMICA.

Throbbing, gnawing pain, aggravated by eating, or exposure to the open air.

3. PULSATILLA.

Shooting pain, that extends to the ear of the affected side, increased by warmth and by rest; mitigated by cold; paleness of face.

4. CHAMOMILLA.

Drawing or jerking pain, attended by heat, and redness of one side of face; aggravated by eating any thing, whether hot or cold.

Remedies for consultation : —

> SULPHUR; BELLADONNA; BRYONIA; RHUS; ARSENICUM.

Administration of Medicine. Three globules may be taken dry or dissolved in three spoonfuls of water and after the first dose, should the pain be greatly increased, it must be regarded as a medicinal aggravation, and the dose should not be repeated, for relief will soon follow. If a recurrence of pain takes place after an interval of improvement, the medicine that afforded relief should be repeated, the repetition to be regulated by the degree of improvement. *

---

* It may be well to state here,— and there may be occasion to repeat the statement in relation to other affections, — that although there should be instances of ineffectual treatment through an insufficiency of, or obscurity in, the indications presented, yet a large majority of cases will be fully met, the probabilities of affording relief being decidedly greater than if the prescriber's attention were distracted by a multitude of symptoms common to a large number of different medicines. And the probabilities above alluded to are vastly increased when compared with those attending

TIC DOULOUREUX. (*Facial Neuralgia.*)

This term is applied to pain in the portions of the nerve that is distributed to the face. The sharp, lancinating character of the pain, its short paroxysms, its spreading along the diverging branches of the facial nerve, the absence of inflammation or swelling, distinguish it from the common toothache. That a comparison may be made with the latter, for which it is liable to be mistaken, the "tic douloureux" is inserted here, while it will again be referred to under its proper head of "Neuralgia." It is an obstinate affection, and one which occasionally admits only of palliative treatment. The chief remedies have been found to be, —

1. ACONITE.

For febrile action, local or general; pain, lancinating, excruciating; heat and redness of the face; constant agitation.

2. BELLADONNA.

Pain excited by touching the affected part, or by moving the jaws; congestion of blood to the head and face; convulsive movements of the eyelids, and of the muscles of the face.

3. ARSENICUM

A burning pain with an occasional sensation of cold

---

the use of any of the innumerable empirical remedies, that readily spring up into public credit yearly, to be as readily relinquished in favor of others, equally undeserving, that succeed them.

occurring periodically ; great debility ; pain increased during rest, and mitigated by the external applica· tion of heat.

Remedies for consultation : —

NUX VOMICA; COLOCYNTH; LYCOPODIUM; PHOSPHORUS.

Administration. Three globules should be dissolved in one table-spoonful of water, and a tea-spoonful given at a dose. If relief or aggravation of pain follows the first dose, it should not be repeated for three hours, and not then if improvement seems to progress. Should there be no effect produced, a second dose may be given in half an hour. If not effectual, another medicine should be then selected.

APHTHA. (*Thrush.*)

In this disease, the membrane that lines the gums, tongue, and palate is affected by little white vesicles, which present the appearance of ulcerated spots, resembling small particles of curdled milk. It chiefly attacks children, sometimes adults, and when slight, is confined to the interior of the mouth, and is easily removed ; but if severe and long continued, the small ulcers line the mucous membrane as far as the stomach, and even beyond it. The formation of aphthæ is frequently attended with difficulty of breathing, want of sleep and appetite, restlessness, dryness and heat in mouth and throat, weakness, and stupidity. The principal medicines in use are, —

1. MERCURIUS VIVUS.

For unusual secretion of saliva, the thrush presenting the appearance of ulceration.

2. SULPHUR.

For same symptoms as above described, should *Mercurius* not prove beneficial, from previous abuse of the medicine or other causes.

3. ARSENICUM.

A blue, livid appearance of the spots, with debility and diarrhœa.

For reference : —

NUX VOMICA; PULSATILLA.

DOSE. Three globules, dry, night and morning, to be continued for three days.

A weak solution of *Salts of Borax* may be applied with a small brush to the ulcerated spots, if the medicine given proves inefficacious. Frequent bathing should be resorted to, with care in feeding, as inattention to cleanliness is sometimes the exciting cause of this affection.

MUMPS. (*Parotitis.*)

An inflamed, painful enlargement of the gland that is situated under the ear, known by the name of "parotid gland," constitutes what is commonly called "mumps." The swelling gradually spreads upward to the cheek and down below the jaw, increasing until the fourth day, when it slowly subsides. When the enlargement is suddenly checked by external applications, or by exposure to cold, the inflammation is sometimes transferred to other locations, where it be comes more troublesome. The chief and only medicine in ordinary cases is

MERCURIUS VIVUS.

Should a high degree of redness, resembling *Erysip-elas*, appear, it will be well to make use of

BELLADONNA.

Both of the above may be taken in dry globules, four at a dose, morning, noon, and night, or the same may be dissolved in four tea-spoonfuls of water, and one fourth given every third hour.

INFLAMMATORY SORE THROAT. (*Quinsy.*)

On looking into the mouth, there are to be seen at the base of the tongue, on each side of the double-arched veil of the palate, almond-shaped glandular bodies, denominated "tonsils." In quinsy these are the seat of the inflammation that extends over the contiguous parts, rendering breathing and swallowing difficult and painful. The constitutional disturbance caused by this inflammation is frequently considerable. The disease may often be arrested in its first stages, but should the means used for this purpose be ineffectual, the swelling of the tonsils will increase, suppuration, or the formation of an abscess, will ensue, and no permanent relief can be experienced until the bursting of the abscess, which occurs on the sixth or seventh day. When one tonsil is healed, it not unfrequently happens that the opposite one will become inflamed, and pass through the same course.

Nothing more than warm water as a "gargle," or the inhalation of warm vapor, both of which promote suppuration, should be used while taking the medicines. From the first to the fourth day, the application of a cloth dipped in cold water around the throat, covered completely with a dry bandage, will be a useful auxiliary in subduing the inflammation; but after this period, if the inflammatory symptoms continue unmitigated, suppuration will almost invariably take place, and warm water should then be substituted for the cold, as the latter delays while the former promotes suppuration.

The following are the chief remedies after the administration of *Aconite*, should much febrile action be present, viz. : —

1. **BELLADONNA.**

Shooting pain when swallowing; sense of contraction about the throat; thirst; accumulation of phlegm in throat; headache.

2. **MERCURIUS.**

Swelling of the muscles of the throat; chilliness; perspiration; pain in the ears.

One of these two remedies will in most cases be sufficient to control the distemper. Either may be given at intervals of two, three, or four hours, according to the urgency of the symptoms. It will be very seldom necessary to resort to any other medicine, but at times one of the following may be indicated : —

CHAMOMILLA; IGNATIA; NUX VOMICA; LACHESIS; PULSATILLA.

DOSE. — Ten globules to be dissolved in two table spoonfuls of water, and one tea-spoonful given every second hour.

DIET. — The food should be mild and unirritating, as arrowroot, gruel, sago, tapioca, &c. No animal food ought to be taken.

ULCERATED SORE THROAT.

This is a more serious disease than the preceding, and is distinguished from it by the white specks which appear in the throat, and the accompanying debility, with a small, fluttering pulse.

The inflammatory sore throat attacks those of a full, plethoric habit, and prevails in cold climates, while the ulcerated sore throat occurs chiefly in warm countries and seasons, and affects feeble children or adults.

The latter is also contagious, which is not the case with the former. The ulcerated spots are seen in the throat soon after the commencement of the inflamma- tion, and there is a dark red or livid appearance of the membrane lining the mouth and throat, as well as of the external surface.

The prostration of strength is very great, and there appears, if the attack is severe, — then termed " malignant," — a scarlet efflorescence on the skin, or large spots of a dark red color. The following should be given : —

1. BELLADONNA; 2. MERCURIUS;

In alternation, every two hours.

Although the two remedies above are equally as effectual as in the inflammatory species, when the disease is mild, yet in severe cases, when from the beginning the debility is extreme, and there occurs vomiting, with small, irregular pulse, parched tongue, and indications of great prostration, another medicine must be given, which is *Arsenicum.* It may be necessary to consult the article on "Typhoid Fever," as similar symptoms to those there described are occasionally here manifested.

DOSE and DIET as in the preceding.

DYSPEPSIA. (*Indigestion.*)

There are numerous disordered states of the stomach, arising from a variety of causes, to which the above term is applied, and the difficulty and absurdity of prescribing for a name is in no instance more evident than in this complaint. Whenever the process of digestion becomes irregular and unhealthy, a group of symptoms are always manifested, to which the characteristic properties of certain vegetable or mineral substances that are medicinal correspond. Whatever may be the proximate cause of the peculiar nature of the derangement of digestion, if in the stomach, there almost always exist prominent symptoms common to all its morbid conditions, pointing. to a few remedies only, which will be designated below in order, with their particular indications.

1. NUX VOMICA.

For sense of fulness and tenderness in stomach; bitter or acid taste; thirst; nausea; constipation; headache; irritability.

2. PULSATILLA.

Pain in stomach; want of thirst; diarrhœa; giddiness; mildness of temper.

3. CHAMOMILLA.

Sensation of sinking in stomach; oppressive pain in region of heart; vomiting of food; shooting pain in temples; excessive thirst.

4. BRYONIA.

Burning at pit of stomach, especially on moving; chilliness; aversion to food; hiccough; constipation.

The ordinary symptoms of dyspepsia, which include those common to all the above remedies, are not detailed, since the first medicine, here as elsewhere, is to be preferred unhesitatingly, as best adapted to the removal of the disease under consideration, unless strikingly marked indications attract attention to another.

Medicines for consultation: —

IPECACUANHA; IGNATIA; SULPHUR; CINCHONA.

DOSE. — Four globules are to be taken, dry, morning and evening.

DIET. — Without a systematic regulation of the diet, as to the period of meals, the quantity and quality of the food taken, no medicine of any kind will prove permanently beneficial to one suffering from indigestion. It is a disorder induced principally by intemperance in eating and drinking, by an abuse of tea and coffee, the use of artificial stimulants, whether drank in the different alcoholic preparations, or eaten in the form of spices, and, above all, by the almost exclusively American habit of swallowing every description of food without proper mastication. It is wholly in vain for any one to expect a cure by medicine of any of the different forms of indigestion, unless strict and persevering temperance is practised.

A pint and a half of gruel, in three equally divided portions, has been recommended as the best daily food for the dyspeptic ; and when improvement results from this temperate course, a return to a more nourishing and solid aliment should be gradual. Whatever produces uneasiness and pain should be avoided. White flour bread is unsuitable food, especially if warm and new. Gravies, acids, spices of every description, must be discarded from the table. Cold water should be the only drink. For further particulars, the directions in the dietary table may be consulted, the reader bearing in mind that no article in the prohibitory list comes under the class of simple food. To rigid and patient abstemiousness in eating and drinking should be added daily exercise and ablution.

INFLAMMATION OF THE STOMACH. (*Gastritis.*)

This severe affection consists of inflammation, confined to the inner coating of the stomach, and is induced principally by irritating substances which have been swallowed. Cold water, drank while the system is in a heated state, is one of its most frequent causes. It has been known, also, to result from external contusion, and from a sudden transfer of inflammation from other parts. Emetics have produced this inflammation. It is characterized by constant and severe pain in the stomach, with a sensation of heat and fulness, great distress, with cold extremities, thirst, a frequent, hard pulse, extreme weakness, spasms, and vomiting after any thing is swallowed, whether solid or liquid.

Aconite is always to be given first, in this as in all other diseases, where much fever is present.

1. ARNICA.

When caused by a blow or fall; pain like that from a bruise.

2. PULSATILLA.

When obstructed menstruation is the cause.

3. ARSENICUM.

When there is great prostration of strength.

DOSE. — Three globules, in a dry state, every half hour.

DIET. — Gum arabic water, barley water, arrowroot or flour gruel, should be continued for two or three days after all traces of this inflammation have disappeared.*

SPASM OR CRAMP OF THE STOMACH. (*Cardialgia.*)

This is a contracting or gnawing pain in the stomach, extending to the back, attended by faintness, nausea, and sometimes vomiting, cold extremities, with anxiety, and is usually induced by the presence of indigestible food, or by the use of tea, coffee, and alcoholic stimulants. The paroxysms are connected with an unnatural condition of the nerves of the stomach, and they are liable to recur so long as care is not exercised in the selection of food.

1. NUX VOMICA.

Pressing pain in stomach after a meal, or the use of coffee and stimulants; nausea.

---

* The above is a serious complaint, often requiring for its proper man agement the skill of a well-instructed physician. All its consequences and complications are not referred to, since the special object of this work would remain unaccomplished were that simplicity and conciseness which prompt domestic administration needs, to be sacrificed to the interminable and perplexing details of a professional treatise. No record, comprising a wide range of diseases, can or ought to be made to supercede in all circum- stances medical attendance.

It is to be presumed that in those of a serious character, a physician's services will, if possible, be obtained. Domestic treatises are to be exclu sively trusted only in so far as relates to the treatment of mild, uncompli cated disorders.

2. PULSATILLA.

When the above symptoms result from menstrual irregularities.

3. BRYONIA.

When the above symptoms, attended with pain in the temples, are increased by moving.

Remedies for consultation : —

CARBO VEGETABILIS ; CHAMOMILLA ; CINCHONA.

DOSE. — Dissolve ten globules in ten spoonfuls of water, and give a tea-spoonful every half hour, until relieved.

DIET. — Oils, cheese, uncooked vegetables, nuts, and stimulants of every description, are to be avoided.

VOMITING OF BLOOD. (*Hæmatemesis.*)

The exciting cause of this hemorrhage is sometimes the presence of a highly irritating stimulant in the stomach, or an external bruise; and it is often symptomatic of some disturbance in other near or remote organs, as obstructions in the liver, spleen, &c., or menstrual irregularities. The blood which proceeds from the stomach may be distinguished from that which comes from the lungs by its color, it being darker and thicker when from the former than when from the latter; and it is in larger quantity, being vomited rather than expectorated, as in hæmoptysis. Vomiting of blood is also usually preceded by a sense of weight, pain, or anxiety in the region of the stomach, and is not accompanied by a cough, as is always the case when the blood proceeds from the lungs.

1. ARSENICUM; 2. IPECACUANHA;

To be given in alternation.

These two medicines will in most instances soon arrest the bleeding ; but in case their operation should not be promptly beneficial, the application of a mustard poultice over the region of the stomach may, by diverting in some degree the irritation to the surface, assist the medicinal curative action.

DOSE. — Ten globules should be dissolved in two table spoonfuls of water, and one tea spoonful given every fifteen minutes.

DIET. — The food should be light and easy of digestion, for a few days subsequent to an attack, and more nourishing as the person is more exhausted.

COLIC.

This term is commonly applied indiscriminately to every pain in the abdomen ; but colic arises from various causes, and is attended by different manifestations.

It may be produced by irritation of bile, by flatulence, by the absorption of lead, by the presence of acrid substances, or by inflammation. In the latter case, to which more particular reference will be made hereafter, the pain is equable, and fixed in one spot, and there are present symptoms of fever, as specially indicated by a rapid pulse. Pressure also increases the pain of inflammation, while it relieves that arising from other causes

Under all circumstances, the following medicines are the most frequently applicable : —

1. COLOCYNTH.

Extremely acute, cramp-like, cutting pain, with tense abdomen, and great restlessness.

2. NUX VOMICA.

Pressing and burning pain, attended with nausea or vomiting.

Reference : —

PULSATILLA; BELLADONNA.

DOSE. — Six globules are to be dissolved in six spoonfuls of water, and one spoonful given every fifteen minutes, the interval between the doses being lengthened to an hour or more, as the pain decreases. (See article on "Enteritis," if the pain is aggravated by external pressure.)

Fomentations, by cloths dipped in warm water, and covered with a woollen bandage, applied over the seat of the pain, and renewed, so that the warmth may be unmitigated and constant, will conduce to relief, in all cases that are not attended by excessive tenderness to pressure.

HEARTBURN. WATER BRASH. (*Pyrosis.*)

By the above term is signified a sensation of acrid heat in the region of the stomach, affecting also the throat; and the word is wholly inapplicable as referring to the heart for its seat. It arises from some irritating cause in the stomach, as spices, and aromatics, strong stimulants, tea, sharp acids, the presence of bile, &c., and is attended with nausea and oppression, sometimes pain and swelling.   The principal remedy is

CHAMOMILLA.

Six globules, taken in a dry state.

If this proves insufficient, one drop of sulphuric acid, taken in a table-spoonful of water, will seldom, if ever, fail to relieve.

CONSTIPATION.

This condition is not always idiopathic, or independent of other disorders, but is owing to the withdrawal of vital energy, occupied in developing disease elsewhere.   It is frequently associated with chronic disturbances.   When not symptomatic, it may be easily removed by medicine, in conjunction with exercise and a suitable diet.   It is now almost universally acknowl-edged that purgatives, so generally and constantly resorted to in former times, have been of far more injury than benefit, augmenting the difficulty that they were employed to remove.   The vast variety of cathartic combinations, the numerous advertised specifics for the cure of costiveness, amply demonstrate the unsatisfactory nature of this system of drugging.   The reaction from the operation of drastic pills, manufactured by ignoble quacks, is invariably increased torpor of the intestines, requiring a daily-augmented quantity of medicine, until the enormous amount necessary to produce the desired effect deranges the whole function of digestion, and ultimately produces disease that is absolutely incurable. The constipation that often attends convalescence from fevers, the comparative disregard of which, among homœopathic physicians, excited alarm in the minds of those first subjected to this mode of treatment, is the natural result of a general disturbance of the secretions, and as improvement progresses, an equilibrium is being reëstablished, and the intestines are resuming their healthy action. Purgatives do not hasten this recovery, but, on the contrary, retard it by diminishing the strength which nature is struggling to restore. The medicines most proper to assist in reëstablishing the natural action, are the following : —

1. NUX VOMICA.

For constipation, caused by indigestible food, sedentary habits, or the use of stimulants ; headache ; loss of appetite ; hemorrhoids ; irritability.

2. BRYONIA.

For the same symptoms, with acidity of stomach, and heat in forehead, in case of failure of *Nux Vomica*.

3. PULSATILLA.

Similar symptoms, particularly if manifestly caused by oily food, and chilliness, with melancholy, is ' present.

4. OPIUM.

Pulsation, and sense of weight in abdomen ; rush of blood to the head ; feeling of constriction in lower intestine.

Reference : —

SULPHUR; LYCOPODIUM; SEPIA; VERATRUM.

DOSE. — Four dry globules, morning and evening.

DIET. — A moderate use of fresh and dried fruit, Indian or rye meal, instead of wheat flour. The avoidance of highly-seasoned food, and all kinds of pastry. A glass of cold water, drank every morning before breakfast; bodily and daily activity, vigorous walks in the fresh air; are the most important of the means to be adopted for the removal of the above complaint.

DIARRHŒA.

This disease depends on irritation, by some cause excited, that increases the ordinary expulsive movement of the intestines, or the secretion from their lining membrane. Many influences give rise to it, the chief of which are sudden atmospheric changes, indigestible food, crude vegetables, unripe fruit, checked perspiration from exposure to cold, great fatigue, and sometimes mental emotion. The same causes, it is well known, do not operate on all alike. Some individuals, for example, will be able to eat with impunity certain articles of diet that are in general directly productive of diarrhœa. The complaint is usually attended with some degree of pain, thirst, impaired appetite, and tenderness on pressure. The mixtures frequently given to check it often add to the existing irritation, and, if efficient as astringents, arrest an in-

DISEASES OF THE DIGESTIVE FUNCTION. 67

creased secretion which nature may have temporarily established for a salutary purpose.

There are various homœopathic remedies for this disorder, according with the variety of causes producing and circumstances attending it. But those adapted to the common description of diarrhœa are, —

1. DULCAMARA.

Especially when the result of cold.

2. PULSATILLA.

When resulting from errors in diet.

3. CHAMOMILLA.

For children; bilious diarrhœa, attended with much pain, furred tongue, thirst, and want of appetite.

4. SULPHUR.

In obstinate cases, worse at night, or when accompanied by eruptions on the skin.

Reference :—

MERCURIUS; BRYONIA: ARSENICUM; CINCHONA.

DOSE. — Six globules are to be dissolved in six tea spoonfuls of water, and one tea-spoonful given after every operation.

DIET. — It is very important that the patient should be as quiet as possible, be suitably clothed, and be particularly careful with regard to food. Exercise should not be taken, even when the disease is mild in character. The surface of the body, particularly the lower part, should be protected by flannel clothing. The diet ought to consist of toasted bread and tea, rice, sago, tapioca, and mild mucilaginous preparations, the quantity of food taken being less than usual. Eggs, milk, acids, fruit, vegetables, meat, fish, beer, and coffee must not be used. Even after the disease appears to have been subdued, it is proper to adhere for a few days to this diet and regimen, as sudden exposure to cold, especially at night, a fatiguing walk or ride, intemperate eating or drinking, will tend to bring on a relapse, or convert the disorder, which is perfectly manageable with care, into the more serious affection that is now to be described.

DYSENTERY.

The characteristic symptom of dysentery is what is termed "tenesmus," a sensation of constriction in the lower portion of the intestinal canal, attended with heat and forcing pains, inducing efforts that prove ineffectual to remove the seeming cause of the pain. In fact, this inclination is altogether independent of the existence of any substance that requires removal. The passage of blood is another peculiar symptom. A continued fever attends the disease, which, as is usual in febrile diseases, increases on the approach of night; and with the increase of fever the local symptoms are aggravated. The exciting causes are like those of diarrhœa, viz.: checked perspiration, local irritation, improper food, exposure to dampness and cold, especially at night. Low, marshy locations favor its production. This severely painful disease is not unfrequently

faial in our country, particularly during the autumn,
when cold days and nights succeed the excessive heat
of summer, and it often prevails as a malignant epidemic. It is to be supposed, of course, that in its
severer forms it will not be left to domestic treatment,
if a physician's services are to be procured.

For the accompanying fever, *Aconite*, as in every
case of fever, should always be first given, six globules
of which are to be dissolved in six spoonfuls of water,
and one spoonful given every two hours during the
first day of the attack. For the disease when fully
developed the following medicines are indicated : —

1. MERCURIUS COR.; 2. COLOCYNTH;

In alternation every two hours.

In the cases of children : —

1. IPECACUANHA; 2. BRYONIA;

In alternation, as above.

DOSE. — Dissolve ten globules of each medicine **in**
ten spoonfuls of water, and give a spoonful first from
the water containing the *mercury*, and in two hours
after, a spoonful of the *colocynth;* after the lapse of
two hours more, if the symptoms continue unabated in
severity, give again of *mercury*, to be again followed
by the *colocynth*, and so on, alternately repeating the
medicines until there is a decided improvement, and
proceed in the same way with the others.*

* Whenever the word " spoonful " is mentioned, the tea-spoon is **referred**
to, unless otherwise expressed.

DIET. — Nothing solid should be taken while the disease lasts, the diet consisting solely of such fluids as are mucilaginous, like gum arabic or barley water, flaxseed tea, slippery elm bark, and thin arrowroot, these articles being sufficiently nutritious, while at the same time serving the desirable purpose of lubricating the inflamed surface to which the disease is confined. A return to more solid food must be gradual, chicken broth, or beef tea being given after improvement commences. It is necessary that the abdomen be protected from cold by a flannel dress constantly worn next the skin. Warm fomentations are often of service in allaying pain.

CHOLERA.

There are two diseases now described under this term, viz., "cholera morbus," an old, prevalent disorder of warm seasons, and "Asiatic cholera," that is of more recent origin, and of much more serious character.

The "cholera morbus" — for a long time the only species known under the new generic term of cholera — consists of vomiting and purging of bile, attended with violent pain, great thirst, and anxiety. In the worst cases there are spasms, cold extremities, and rapid respiration with a very feeble pulse. During the month of August, it prevails the most frequently in our climate, and its severity is generally in proportion to the heat of the weather. It is produced by indigestible, acrid, irritating food, and rarely occurs spontaneously, or without an evidently direct cause.

The disease may often be controlled in its first stages by

1. CHAMOMILLA;- 2. IPECACUANHA;

Given in alternation.

In the severer form by

1. VERATRUM; 2. ARSENICUM;

In alternation.

DOSE. — The above remedies are to be used in water, Ten globules dissolved in ten spoonfuls of water, and one spoonful given at a time. The doses of each are to be administered every half hour, or every hour, according to the violence of the symptoms

ASIATIC CHOLERA.

This disease, differing in some particulars from the preceding, is characterized by the white appearance of the fluid ejected by vomiting and purging, by a deficiency of the urinary secretion, by the great coldness and the lividity of the skin, by its rapid progress and fatality.

The lamentable uncertainty and want of success in the treatment of this singular epidemic by the empiric-like " regular " mode of practice, and its comparatively easy control by " infinitesimal doses," present a striking contrast ; and this well-established fact is convincing, were there no other proofs, of the truth of the homœopathic principal of cure. In general, the premonitory symptom is diarrhœa, which, if neglected or improperly treated, will terminate in fully-developed cholera. This should be treated as described in the article on "Diarrhœa." When this symptom is accompanied by others, as nausea, severe pain, giddiness, sense of oppression in stomach, flatulence, headache, no time should be lost in giving

CAMPHOR,

In the form of tincture, or a saturated solution.

DOSE. — One drop in a spoonful of cold water, given every five minutes, and repeated at intervals lengthening as the symptoms subside. If the disease does not yield in three hours, and the more violent symptoms of confirmed cholera should appear, give

1. VERATRUM; 2. CUPRUM;

In alternation.

DOSE. — Twelve globules of each of the above are to be dissolved in ten spoonfuls of water, and two spoonfuls from each alternately given every fifteen minutes, half hour, or hour, according to the violence of the symptoms. Should amelioration not soon take place under the administration of the above, the following should be given, in the same doses, and at the same intervals, viz. : —

1. VERATRUM; 2. ARSENICUM;

In alternation.

Two of the above, namely, *Veratrum* and *Cuprum*, are given with much confidence as prophylactics, or preventives of cholera; two or three globules of each, taken alternately every other day, are reported to have preserved all who have adopted this precautionary measure. There are many predisposing causes which should at the same time be avoided, such as sudden or long exposure to cold or dampness, depressing emotion, artificial stimulants, intemperance in eating or drinking, uncooked vegetables and fruit, close confinement in crowded or ill-ventilated rooms, fasting when there is an appetite, and eating when there is none.

DIET. — Cold water is the best drink, in small quantities at a time. Small pieces of ice, allowed to melt slowly in the mouth, will assuage thirst, and is grateful. Nothing more should be given during an attack of cholera. Warm applications may be made to the surface of the body and to the feet, with constant, hard friction with the hands or heated flannel cloths, particularly where there is great coldness, pulselessness (collapse). During convalescence, much care should be exercised with regard to the indulgence of the appetite. When free perspiration occurs after a severe attack, all danger is usually past.

HEMORRHOIDS. (*Piles.*)

This name is applied to enlargements of the veins that exist at the inferior portion and termination of the lower intestine. In the veins here situated, the blood, in its return to the heart, is impeded in its course by resistances that do not prevail elsewhere, and the consequence is distention, from which results inflammation, great pain being thereby caused. The tumefied vessels are sometimes ruptured by the great distention. Whatever is capable of retarding the progress of blood in these veins, denominated "hemorrhoidal," is to be regarded as contributing to the production of this disease. It is generally attended by, and is the effect of, constipation, and the best measures for the prevention and removal of this chief cause of hemorrhoids are to be resorted to. The majority of cases, under whatever form, will be relieved by the following remedies:

1. SULPHUR; 2. NUX VOMICA;

In alternation.

DOSE. — The above are to be taken in dry globules, three at a time, each medicine daily, the first in the morning, the second at night.

References: —

BELLADONNA; RHUS; PULSATILLA.

There are medicines besides the above named to be used under certain circumstances, but in any other than simple, uncomplicated cases, the relations sustained to chronic complaints of a different nature are so intimate, that unprofessional prescribing would be of little use, and could not be properly adopted. Great patience and perseverance, as well as skill, are needed for the permanent removal of chronic and complicated affections of a hemorrhoidal character.

DIET. — All persons subject to the above complaint should live on a light diet, of a cooling nature, avoiding .

coffee, tea, spices, and stimulants ; take often laxative food, as stewed prunes, dishes served up with oil or molasses, and drink much of cold water. The use of purgatives, instead of mitigating, has a direct tendency to aggravate the disorder. Astringent and stimulant preparations, of any sort, should not be used externally. The very best of all local applications is cold water, — the colder the better, — employed frequently.

JAUNDICE. (*Icterus.*)

The immediate cause of this disturbance is an inac tive condition of the liver, on account of which inactiv ity the bile, a peculiar secretion of the liver, instead of taking its natural course, passes into the circulation, and tinges the cutaneous blood vessels with a yellow hue. It may also be the result of a spasmodic stricture, or other obstruction of the canal or duct through which the bile flows from the liver. The chief predisposing causes are, a sedentary life, strong mental emotion, especially anger, artificial stimulants, and the free use of purgative medicines. It is distinguished principally by yellowness of the eyes, and subsequently of the whole body, a weak pulse, languor, bitter or acrid taste, dry heat of the skin, and constipation.

1. MERCURIUS,

To be followed, if necessary, by

2. CHAMOMILLA,

When directly caused by passion.

3. NUX VOMICA,

When caused by inactive life, and the use of **stimulants.**

References : —

CINCHONA; PULSATILLA; BRYONIA.

DOSE. — Six globules, taken dry, and repeated every **five hours.**

DIET.—The food should consist of fruit, vegetables, and but little meat, of any kind. Oily articles, stimulating drinks, coffee, tea, wine, and beer, and spices **of all kinds,** should be strictly avoided.

INFLAMMATION OF THE LIVER. (*Hepatitis.*)

There are many disturbances, which have their origin and principal seat in the stomach, that are erroneously termed " liver complaints." While inappropriately used to designate the character of many disorders of digestion, the term is often, to a certain extent, correct, as the liver is more or less affected in most derangements of the stomach. The liver, however, is subject, with other organs, to acute and chronic inflammation, the former being characterized by symptoms of fever, with a dull pain, sometimes sharp, in the right side, extending to the right shoulder ; difficulty of breathing, cough, and vomiting : the latter by constipation, loss of appetite and flesh, indigestion, yellowness of skin, but without fever — symptoms like those described as attending jaundice. In both forms of inflammation, the surface over the region of the liver is hot, swollen, and sensitive to pressure. The yellowness of skin is owing to a disturbed secretion of the bile, which is, on this account, retained in the blood. Affections of the liver are very common in warm climates. Heat induces them, as well as mental emotion, the use of stimulating food and drink, and medicines. In the acute form of this inflammation, *Aconite* should always be given, as often described, for the fever at the commencement, and afterwards.

1. BELLADONNA; 2. BRYONIA;

In alternation.

Should there be pain, heat, and swelling in the region of the liver, after the active symptoms have been subdued, or should it appear in the chronic form, the chief remedies are, —

1. PHOSPHORUS; 2. BRYONIA;

In alternation.

Reference : —

NUX VOMICA; ARSENICUM; SULPHUR.

DOSE. — Six globules of each medicine are to be dissolved in six spoonfuls of water, and one spoonful is to be given from two to four hours, according to the severity of the symptoms.

DIET. — The food should be of the same mild, unirritating character as in other inflammatory diseases — gruels, arrowroot, and similar preparations, with a gradual return to a more solid diet as the fever subsides.

7 *

INFLAMMATION OF THE SPLEEN. (*Splenitis.*)

Like the liver, the spleen is subject both to acute and chronic inflammation, the first form of which comes on by a sudden attack of shivering, followed by heat and thirst, with shooting pain in the left side, opposite the lower rib, and swelling in that region. It occurs less frequently even than inflammation of the liver in our climate, and the causes seem to be common to both. Persons of a full, plethoric habit are more subject to its attacks.

The medicines principally indicated are, —

1. CINCHONA: 2. ARSENICUM;

In alternation.

Reference : —

NUX VOMICA; ARNICA; BRYONIA.

DOSE, DIET. — Same as for the preceding.

INFLAMMATION OF THE BOWELS. (*Enteritis.*)

There are two varieties of this inflammation, one in which the intestines themselves are inflamed, and the other where the inflammation affects the abdominal membranes enclosing them. The symptoms common to both, are severe pain in abdomen, feverish symptoms, obstinate constipation, small, frequent pulse, and extreme sensitiveness to pressure on abdomen, the weight even of the bed clothes being insupportable. Among the chief causes are exposure to cold, indigestible food, and long-continued constipation. The principal medicines, after the use of *Aconite*, in solution, under the circumstances repeatedly referred to hitherto, are, —

1. ARSENICUM; 2. VERATRUM:

In alternation.

Dose. — Dissolve ten globules in four table-spoonfuls of water, and give one tea-spoonful every hour, lengthening the interval as improvement progresses.

Diet. — A strict confinement to the most simple food and drink is very important; nothing solid or stimulating should be used.*

WORMS.

Many varieties of these insects are found in the human intestines, the principal of which are known under the appellation of the thread worm, the round worm, and the tape worm. It will serve no important practical purpose to enter into a particular description of all the varieties, and of the theories relative to their origin, &c. It is sufficient to state that the general symptoms by which their presence in troublesome numbers is indicated are, a voracious appetite, disturbed sleep, feverishness, leanness, difficult breathing, itching of the nose, pale countenance, offensive breath, and cough. The rational method of cure is to remove

---

* This is one of those serious complaints which it is proper to introduce here in order to the completeness of this class of diseases, but under the presumption that it will not be subjected to domestic treatment except under the most pressing necessity.

the morbid condition that furnishes the unhealthy secretions in which the worms live. This is not to be
accomplished by violent expellents, since they often
accumulate as fast as they are removed, but by cor-
recting the secretions referred to, and by strict attention to diet. The best direct remedial measures for
their removal are the use of salt-water injections, and
a patient administration of the following, viz. : —

1. SULPHUR; 2. MERCURIUS;

In alternation.

Dose. — Three globules of each, morning and evening. If not soon successful, give

8. SANTONINE.

Dose. — Half a grain twice in 48 hours.

Diet. — Ripe, uncooked apples may be freely given,
with dry, toasted bread. A free use of salt in food is
recommended. Sweetmeats, candies, cake, and pastry
should on no account be allowed.

CHAPTER II.

DISEASES OF THE ORGANS OF RESPIRATION.

ACCORDING to the plan proposed of introducing the different classes of disease by a brief description of the healthy organs and their functions, we commence this chapter on the respiratory system, by reference to the larynx, or entrance to the lungs.

At the root of the tongue, directly in front of the passage leading to the stomach, is a tube, or canal, known as the windpipe, composed of a series of cartilaginous rings, extending from the opening called the glottis to a point not far from the top of the breast bone. The tube there divides into two smaller canals, that diverge from each other, one passing to the left, the other to the right lung. As each of these bronchial tubes enters the lungs, it divides into numerous smaller tubes, that spread out from each other like the branches of a tree, gradually decreasing in size. Anatomists designate the upper third of the air passage " the larynx," the middle portion " the trachea," and the diverging portions the " bronchii " or " bronchial tubes." The glottis, or superior portion of the larynx, is surmounted by an oval cartilage, which is always elevated by its own elasticity, and closed only when food is forced by the act of swallowing over and beyond it into the passage leading to the stomach. This covering is called the " epiglottis," and when the act of speaking and that of swallowing are attempted at the same time, it is elevated in order to afford passage for air into the larynx, the vocal organ; and consequently the food, not being able to pass over it, falls into the larynx, and produces choking, terminating in death by suffocation, should the food not be soon removed by coughing, or by some other means.

The lungs are two cone-shaped bodies, situated on each side of the chest, and separated from each other by the heart, and by a membranous partition. The lung on the right side is divided by deep fissures into three parts, or lobes, and is the largest. The left lung has but two lobes. They are surrounded by a delicately thin membrane, called " pleura," the continuation of that which separates them. This is subject to the inflammation known as pleurisy.

Inside, the lungs are filled with cells containing air conveyed through the innumerable ramifications from the main bronchial tubes. These cells are alternately filled with, and emptied of, air by the process of breathing. Blood is sent to the lungs through two large arterial vessels, which vessels, after being divided and subdivided, are distributed upon the sides of the air cells in the form of delicate network, and through the extremely thin membrane which forms the cells, every particle of blood is exposed to the air inhaled, being thus purified and rendered fit for circulation. By this contact with the air in the lungs, the blood is changed from a dark purple to a bright scarlet color, having absorbed all the oxygen, or life-giving principle of the air, and parted with an equal quantity of carbonized or fixed air, which it had received in circulating through the system. So important is this constant renovation of the blood that the least interruption of breathing is seriously felt, and the process cannot be entirely suspended, even for a very short time, without destroy. ing life.

The natural warmth of the body originates from, and is preserved by, the union in the lungs of the carbon in the blood with the oxygen in the air; and the operation is like the combustion of wood or a candle, the heat producing flame from which is the joint product of carbon and oxygen.

When portions of the lungs are diseased, a free circulation of air and of blood is interrupted, and the general health is impaired. Whatever tends to diminish the capacity of the lungs, or obstruct the free entrance of air into them, or to prevent the full expansion of the chest, disturbs the perfect action of the important function of breathing. There are differences in different individuals as to the size of the lungs, the capacity of their blood vessels, the freedom of circulation in them, and the nervous energy devoted to them, while their condition varies under the operation of all healthy and unhealthy influences.

This chapter will be devoted to the consideration of those disorders affecting all the organs connected with the respiratory system.   · ·

CATARRH. (*Cold in the Head.*)

This term is applied to an inflammation of the mucous· membrane that lines the nasal passages, as well as the interior of the mouth and throat. There are two species of catarrh, one of which is very common, known as "cold in the head," the other as "influenza," or epidemic catarrh. The first or common catarrh commences in the nose, the irritation generally extending down the throat, towards the lungs. It consists of an increased secretion from the mucous membrane, more or less irritating, sneezing, loss of smell, free secretion alternating with obstruction, and if extending to throat, cough, with feverish symptoms.

Its simplest form, that of cold in the head, will often be removed alone by, —

1. NUX VOMICA; 2. PULSATILLA;

If the first is unsuccessful.

Other remedies, to be afterwards used if necessary, are, —

CHAMOMILLA; ARSENICUM; MERCURIUS.

When the inflammation has extended to the throat and even the lungs, being attended with fever, with occasional chills, pain in head, back, and limbs, and want of appetite, give, —

1. ACONITE; 2. BELLADONNA;

In alternation.

DOSE. — Six globules are to be dissolved in six spoonfuls of water, and one spoonful taken every three to six hours, according to the severity of symptoms.

DIET. — Solid animal food and all stimulants are to be avoided, and the diet confined to broths and vegetables. The promotion of perspiration by warm drinks taken immediately on going to bed will often assist in the removal of catarrh.

INFLUENZA.

When catarrh attacks the throat and lungs, and in addition to the symptoms described as belonging to catarrhal fever, is accompanied by extreme bodily and mental prostration, it is called "influenza;" by the French the "grippe." It prevails as an epidemic, or disease that attacks many persons at the same time, and is supposed to be owing to some peculiar atmospherical condition. It is intermediate in its nature between a common catarrh and a lung fever, being an inflammation of all the air passages, causing hoarse-

8

ness and wheezing, nausea and headache; and the
sympathetic general fever frequently proves quite violent. To subdue this, *Aconite* is always first to be
given. But for the true influenza, which is characterized by the very great prostration of strength above
mentioned, the true remedy is, —

ARSENICUM.

When an affection of the throat is a predominant
symptom, with swelling of glands, and great accumulation of mucus, a condition which occasionally prevails,
the remedies are, —

1. BELLADONNA; 2. MERCURIUS;

In alternation.

DOSE. — Ten globules in six spoonfuls of water, and
one spoonful given every three hours.

DIET. — Like that recommended in the preceding article on " Catarrh."

COUGH.

This occurs, in frequent instances, as a symptom of
some other complaint, and a great number of affections
induce it, some of which are very remote from the immediate location of the cough. It also occurs as a distinct affection, of a spasmodic character, arising from
local obstruction or irritation in the respiratory passages. As the circumstances attending cough, and all
the varieties of disease of which it is a symptom or
effect, are very numerous, it would not consist with our
plan of simplification to allude to them all. But the

four chief remedies will be found below, with their special indications.

1. NUX VOMICA.

Exhausting cough; bruise-like pain in abdomen; headache, excited by motion and speaking, preceded by dryness and scraping in throat, worse towards morning.

2. PULSATILLA.

Loose cough, with expectoration; nausea; pain in throat and chest; profuse night sweats; loss of appetite.

3. BELLADONNA.

Dry, spasmodic cough, with difficult breathing; constriction in chest, occurring at night; pain beneath the breast bone.

4. IPECACUANHA.

Dry cough, producing nausea and vomiting; constant nasal obstruction, with loss of smell, increased by cold air.

Reference : —

CHAMOMILLA; HYOSCYAMUS; BRYONIA; ARSENICUM.

DOSE. — Six globules, to be dissolved in twelve spoonfuls of water, and one spoonful to be taken every four to eight hours, according to the violence of the cough.

DIET. — All stimulating food and drink should be avoided. Exposure to sudden changes in temperature is, of course, to be guarded against.

WHOOPING COUGH. (*Pertussis.*)

This species of cough, peculiar to children, chara؛ terized by, and named from, the sound produced by drawing in the breath, is of a contagious character, and occurs but once during life. It commences with hoarseness, cough, difficulty of breathing, and febrile symptoms. At the end of ten or twelve days, the peculiar convulsive paroxysms of coughing, attended with a shrill whoop, and terminated by a fit of vomiting, are fully established.

The proximate cause of this cough is believed to be an unnatural secretion of viscid phlegm from the lining membrane of the air passages, adhering so closely as to be with great difficulty detached. When expectoration becomes free, the paroxysms are less frequent, and improvement commences.

During the preliminary stage of the complaint, the treatment consists in the use of those medicines that are indicated by symptoms described in the preceding article. When the characteristic feature, viz., whooping, becomes developed, it is to be controlled by the following remedies : —

1. DROSERA.

When the cough is spasmodic, violent, fatiguing, attended with whistling or whooping inspiration, and terminating in copious expectoration or vomiting; relief when moving.

2. BELLADONNA.

When the cough is worse at night, dry, barking, harsh, and attended with hoarseness and soreness of throat.

3. CUPRUM.

When convulsions occur, followed by loss of consciousness, rattling in chest, wheezing.

4. VERATRUM.

Coughing, followed by great prostration, with extreme thirst and cold perspiration.

DOSE. — Three globules of the medicine selected may be given in a dry state, after each paroxysm of coughing.

DIET. — No animal food should be allowed until after a decided amendment has taken place. The diet must be simple, and exclusively vegetable, as sago, tapioca, gruel, bread, and articles of this nature, until the cough improves and there are no feverish symptoms, when broths may be given, and a gradual return made to the usual diet.

CROUP.

The seat of this disease is in the windpipe, mostly in that portion called the trachea, the lining membrane of which, in consequence of a peculiar inflammation, secretes a thick, tenacious fluid, that coagulates, and, if the disease is not checked, becomes a dense membrane, which, by lessening the diameter of the air passage, obstructs the breathing. It usually attacks children from one to four years old, though it has happened,

8 *

and proved fatal, to adults.   It first manifests itself by
the symptoms of a common cold, by sneezing, and a
cough, with little or no change in the voice or respiration.   The cough soon acquires a shrill, ringing sound,
attended with difficult breathing and a hoarse voice,
the pulse being small and rapid.   On examining the
throat, it is seen to be of a dark red color, and a thick,
glutinous coating may sometimes be seen adhering to
it.   The membrane is sometimes thrown off by expectoration, and in this case the disease terminates favorably.   The immediate cause of a fatal termination is the
obstruction to free respiration, from the presence in the
trachea of this false membrane.   It is not often that
the croup, if proper treatment is adopted at the com
mencement, passes beyond the stage of inflammation.
Aconite is always to be given ; first, for the mitigation
of the feverish symptoms ; afterwards, or should the
croup be at once fully developed, —

1. **HEPAR SULPHURIS ; 2. SPONGIA ;**
    In alternation.

3 **LACHESIS.**
    In obstinate cases, where the above medicines fail,
    or where there is an excessive sensitiveness of the
    throat to touch ; great weakness ; muscular rigidity.

DOSE. — Ten globules of the medicines, to be dissolved in ten spoonfuls of water, and one spoonful
given every hour, the interval to be increased as improvement progresses.   Should the patient grow worse,
instead of better, *Bromine*, first decimal potence, is to
be given every fifteen minutes.*

* A cold-water bandage should envelop the throat from the beginning. ·

DIET. — The diet should consist of barley water, arrowroot, gruel, &c., and if not much fever, broths. All solid animal and vegetable food must be avoided until the disease is wholly subdued.

MILLAR'S ASTHMA.

This affection of the throat is a species of croup, of a spasmodic nature, consisting in a temporary closure by spasm of the fissure of the glottis, or the entrance of the windpipe. It has, no doubt, frequently been confounded with the true croup, with which it differs by the absence of a membranous formation, and by not being ushered in by symptoms of a cold. It attacks children suddenly, usually in the middle of the night, and is indicated by difficulty of breathing, a hoarse cough, a deep, harsh sound of the voice, great anxiety, and apparent suffocation, all of which symptoms terminate in a few hours, by a fit of vomiting, sneezing, or coughing, and are succeeded by a quiet sleep.

Attacks of a similar kind often follow after intervals of a day or two. The chief remedy is

1. SAMBUCUS,

Which, if ineffectual, may be followed by

2. PULSATILLA.

The dose and diet are the same as in croup, which see above.

INFLAMMATION OF THE LARYNX. (*Laryngitis.*)

This disease also resembles croup, but is almost exclusively confined to the larynx, while the seat of croup
is principally in the trachea, that portion of the air-
passage nearer to the lungs. The symptoms are fever,
labored breathing, hoarse voice, painful sensation of
constriction in the throat, frequent pulse, and an exceeding sensitiveness to touch, all common to croup;
but the inflammation terminates in suppuration, instead
of the membranous formation, and there is not present
the barking cough of croup.

The same remedies have ·been · found applicable to
both affections.

1. ACONITE

For the fever, then

2. HEPAR SULPHURIS and SPONGIA,

In alternation. Afterwards, if necessary,

3. LACHESIS.

HOARSENESS

Is a frequent accompaniment of catarrh, and other
inflammatory affections of the air passages; but it is
sometimes the chief, if not the only indication present
of the existence of a morbid condition of the larynx
or trachea. For this may be given,--

1. HEPAR SULPHURIS; 2. LACHESIS;

In alternation.

Dose. — Ten globules of each to be dissolved in six spoonfuls of water, and one spoonful to be given every six hours.

Diet. — The same as for a cold. No objection can exist to the free use of *Gum Arabic Water*, or simple syrup, as the local irritation of all catarrhal affections is alleviated by the lubricating quality of demulcent, mucilaginous fluids.

BRONCHITIS (*Pulmonary Catarrh*)

Is an inflammation of the inner membrane of the two canals which diverge from the trachea, and ramify through each lung, named "bronchial tubes." The previously described diseases have, as was stated, their seat in the upper and middle portions of the windpipe, while this disease is an affection of the same membrane, as it continues lining the two bifurcations above named, that pass into the lungs. Bronchitis appears in an acute and chronic form. The symptoms of the former are chilliness, difficulty of breathing, attended with a sense of oppression in the chest, hoarseness, fever, violent cough, terminating, after a time, in free expectoration. A wheezing sound accompanies breathing. The causes are the same as those which give rise to catarrh, of which it is, in truth, a variety, and when slight, will yield to the same medicines. But when much fever and oppression exist, the principal remedy at the commencement, which is to be continued until the febrile action is subdued, is, —

1. ACONITE.

After which the chief medicines indicated **are,—**

2. BELLADONNA ; 3. BRYONIA ;

In alternation.

DOSE. — Six globules are to be dissolved in six spoonfuls of water, and one spoonful given every three hours until improvement occurs.

DIET. — The food should be simple and unirritating, as in all affections of the air passages, but particularly so during the existence of fever.

The chronic form of this complaint may follow an acute attack, or arise from irritating substances inhaled into the lungs, or it may be the sequel of improperly treated diseases, especially those of the skin.

The symptoms are like those of the acute form, modified in some respects, particularly in relation to fever. If the disease progresses, difficulty of breathing increases, with lassitude, emaciation, frequent and feeble pulse, exacerbation of fever at noon and evening, night sweats. These symptoms are common with tubercular consumption, from which it is not easy to be distinguished by those inexperienced in the modern modes of examination.*

The medicines most frequently indicated in the

---

* It cannot be too often repeated, that the plan of this work would be frustrated, were all those measures to be here detailed that physicians only are competent, after a long course of study, to pursue; or were the large variety of remedies that might be collected together for every possible emergency allowed to encumber these pages

course of this lingering complaint, when fully established, and which it is only proper to designate for
reference, are, —

SULPHUR; PHOSPHORUS; PULSATILLA; CALCAREA;
STANNUM.

INFLAMMATION OF THE LUNGS. (*Pneumonia.*)

The substance of the lungs themselves is subject to
attacks of inflammation, for the proper treatment of
which medical skill is generally necessary. The symptoms are, difficult, rapid respiration, an acute pain in
the chest, with sensation of weight, a hard, painful
cough, dry at first, afterwards followed by expectoration generally tinged with blood. The beating of the
heart is full and strong, often with palpitation; the
face is flushed, with dark red lips, dry mouth and
tongue; intense thirst; the skin dry and of a burning
heat. The cough is always excited by every attempt
at speech or a full inspiration. The pain is aggravated
by the cough, by a sudden change of position, or by
pressure on the chest. The febrile symptoms at the
commencement indicate, —

1. ACONITE   2. BELLADONNA;

In alternation, administered in successive doses,
every two hours, for the first twenty-four hours.
(See DOSE.)

When the fever has been subdued, in many cases,
Aconite and *Belladonna* will be sufficient to remove

the disease entirely. The next remedies to be used are, —

3. BRYONIA; 4. PHOSPHORUS.

For alternations of chilliness and heat; pains in the back; rusty-looking expectoration; pain increased by moving; sharp, stitching pain in different parts of the chest, sometimes burning; loud, rattling res-iration; pain in stomach; oppressive tightness, as f a band encircling the chest.

References : —

ANTIMONIUM TARTARIUM; SULPHUR; RHUS.

DOSE. — Twelve globules to be dissolved in eight spoonfuls of water, and one spoonful given every two hours, so long as the severe symptoms remain un-changed. In alternating medicines, it will be remem-bered that a spoonful of the medicine numbered as the first must first be given, to be followed in two hours by a spoonful of the second, and so continuing regu-larly while necessary.

DIET. — During the prevalence of fever, the nourish-ment should be wholly liquid, as toast water, barley water, &c. ; great caution being exercised, even after the disease has disappeared, in returning to the usual diet, which return should be gradual, beef tea, chicken broth, &c., serving as the introduction to solid animal food.

PLEURISY. (*Pleuritis.*)

This term is applied to an inflammation of the pleura, the membrane that lines the internal surface of the chest, forms a partition dividing the chest into two cavities, and is reflected over the surface of the lungs. It is occasioned by an exposure to cold, and by other causes that induce inflammatory complaints in general. It commences with an acute pain in the side, increased by respiration, and is accompanied by fever, cough, and the other symptoms described as attendant on inflamed lungs, but particularly indicated by the sharp, stitching pain in the side, which prevents free breathing, together with a cough, without the rust-colored expectoration of pneumonia. Both diseases sometimes exist together, and the treatment is very much the same for each, with the substitution of *Bryonia* for *Belladonna*, in connection with *Aconite.*

1. ACONITE; 2. BRYONIA;

In alternation.

3. SULPHUR.

When fever and pain are moderate, but not wholly subdued, after several doses of the above have been taken.

4. ARNICA

Is often of service for pleuritic pains without fever.

DOSE. — Twelve globules are to be dissolved in six spoonfuls of water, and one spoonful given every hour

9

during the active stage ; at other times, every two or three hours.

DIET. — Same as described in the preceding article on " Lung Fever."

ASTHMA.

The prominent symptoms of asthma are, difficult respiration, causing a distressing sense of suffocation and fulness in the chest, with anxiety. It appears in paroxysms, and has been divided into *dry* and *humid* asthma. The former is so called from the absence of expectoration, by which the latter is characterized. This expectoration, occurring in the humid asthma, becomes generally profuse towards the termination of a paroxysm, and affords great relief. The attacks are usually preceded by languor, flatulence, oppression, heaviness over the eyes, with sickness and restlessness. They frequently occur about midnight, and render a recumbent position intolerable, while every exertion is made by the sufferer to inflate the lungs. In many instances there is present from the first a hard, dry cough, which, after one or two hours, is attended by the expectoration of mucus, sometimes tinged with blood. These attacks vary much in violence and duration. The medicines most to be relied on for the relief of suffering during the paroxysm are, —

IPECACUANHA.   .

When roused from sleep by a sense of constriction

in lungs, with rattling in throat, gasping, pale face, rigid muscles, and cold extremities.

2. ARSENICUM.

When, added to the above symptoms, there is great debility and exhaustion, with cold perspiration, and burning sensation in throat.

The above may be administered in alternation, and if not affording relief, as in the majority of cases they will do, the following should be substituted : —

3. PULSATILLA; 4. NUX VOMICA;

In alternation.

DOSE. — Twelve globules to be dissolved in ten spoonfuls of water, and one spoonful to be given every half hour during the paroxysm.

EXPECTORATION OF BLOOD. (*Hæmoptosis.*)

Preceding an attack of this description of hemorrhage, sensations of tightness in chest are experienced, with difficulty of breathing, anxiety, palpitation, a sweet or salt taste in the mouth, a burning, painful sensation in chest, lassitude, and a rapid pulse. The blood proceeding from the lungs is thin, of a light florid color, and is generally raised by coughing; whereas if from the stomach, it is of a dark color, unaccompanied by cough. It is not an affection so dangerous as is commonly supposed, being often only a slight effusion of blood from the lining membrane of the bronchial tubes, and may be occasioned by any violent exertion in those predisposed to it from a faulty proportion in the vessels of the lungs or the capacity of the chest. The medicines almost invariably successful in checking this bleeding are, —

1. ACONITE; 2. ARNICA;

In alternation.

Should there be, after the hemorrhage, any uncomfortable sensations in the chest, debility, difficulty of breathing, chills, &c., the best medicines are, —

1. CINCHONA; 2. PULSATILLA;

In alternation.

DOSE. — Three globules of each of the above, dissolved in a table-spoonful of cold water; the whole given at a dose every fifteen minutes.

DIET. — No stimulants of any description should be given. A light, cool, vegetable diet is the best. The drinks should be cold, the best of which is cold water. Perfect quiet and silence must be observed while the bleeding continues, and even afterwards, as far as possible, while the consequences of the hemorrhage are in any way manifested.

CONSUMPTION. (*Phthisis Pulmonalis.*)

This complaint, so prevalent, and, alas! so fatal, in our variable climate, is well known, after reaching a certain stage, to be incurable by any medical agent. In its incipient form its progress may be arrested; and should restoration be beyond the reach of human skill, much may be done by homœopathy to mitigate distressing symptoms, and smooth the sufferer's path to the grave. It manifests itself to the superficial observer by emaciation, cough, debility, hectic fever, and purulent expectoration, commencing with a short, dry cough, a difficulty of breathing, especially after any exertion, a sense of oppression at the chest, and languor. As the disease advances, respiration becomes more hurried and difficult, the cough more violent, and attended with expectoration of an opaque, viscid matter, greenish, sometimes tinged with blood. Pain is generally felt in different parts of the chest, and in the side ; a hectic fever, known by its appearance about noon, towards evening, and at midnight, with circumscribed redness of the cheeks, burning heat in the palms of the hands and soles of the feet, with profuse perspiration at night. All these symptoms usually continue to the last, while the mind is hopeful and confident of recovery. Many like accompaniments may exist in other disorders, and there is no sure token of the presence of tubercular disease of the lungs but those obtained by auscultation. There is one symptom, however, which is seldom seen except when tubercles exist, and that is, the expectoration, by coughing, of pus, appearing in the form of globular, wool-like masses, which assume a flat, circular form on a dry surface, and remain separate and distinct from each other, while in water they float heavily at different depths, mostly subsiding to the bottom.

9 *

In the different stages of this lingering disease many medicines are indicated, the most important of which are named below.

1. ACONITE.

For febrile symptoms, frequent congestion to the chest, short cough, expectoration of blood.

2 SULPHUR.

In the incipient stage, for hoarseness, difficult breathing, weakness and burning in chest, and for the stage of purulent expectoration.

3. CALCAREA.

Violent cough, more frequent at evening; expectoration purulent and offensive; great exhaustion; night sweats; urgent thirst.

4. PHOSPHORUS.

Soreness and pain in chest; sense of oppression; evening chills; irritability; a hard cough, excited by speaking, and during night; pair and sensitiveness in throat.

Reference :—

STANNUM; LYCOPODIUM; SEPIA.

Dose. — Six globules at a dose, repeated, according to circumstances, from one to four times in twenty-four hours.

Diet. — The food taken should be nutritious and mild, as milk and its preparations, farinaceous vegetables, fruit, shell fish, Iceland moss boiled in milk, &c This mild system of diet has far more reason and experience in its favor than the stimulating plan recommended by many at the present time. Something, however, depends on the previous habits and constitution; and in certain cases the diet may be full and generous. The chief object is, to sustain the strength without exciting inflammation of the lungs. Gentle exercise ought to be taken daily, and the air respired should be pure, and as equable in temperature as can be found.

CARDITIS. (*Inflammation of the Heart.*)

An inflammation of the heart, or of its surrounding membrane, the pericardium, is indicated by violent pain in the left side of the chest, where this organ is situated, or by a painful sense of pressure in that region, together with rapid and difficult breathing, quick, feeble, irregular pulse, high fever, fits of fainting, and great anxiety. It is a disease of rare occurrence. Inflammation of the lungs presents analogous symptoms, and it is not always easy for a person unaccustomed to disease to distinguish one from the other. In carditis, the pain is confined to one spot; there is more oppression and faintness, less cough, and a horizontal position is more tolerable than in pneumonia. The surface of the body is not so uniformly hot in the former as in the latter, the feet and hands being often cold. Whether the seat of the inflammation, however, be the heart or the lungs, the treatment, at the commencement, is the same; and as, in the old practice, bleeding was once resorted to in both diseases, so in the homœopathic treatment, *Aconite* is indicated by the general febrile disturbance; and none can better estimate the advantage that the latter possesses over the former, than those who have been subjected to both modes of treatment.

When respiration is rapid and laborious, the pulse quick and strong, the skin hot, the pain sharp and piercing, with anxiety and faintness, the medicine to be given is

1. ACONITE.

2. ARSENICUM.

Violent, irregular, and painful beating of the heart; feeble pulse; burning, and soreness in the region of the heart; general debility.

Reference : —

DIGITALIS; BRYONIA; SPIGELIA.

DOSE. — Dissolve ten globules in two table-spoonfuls of water, and give one tea-spoonful every two hours.

ANGINA PECTORIS.

This disease, which occurs in paroxysms, is characterized by an acute, drawing pain in the chest, extending from the lower end of the sternum or breast bone, into the left arm, the pain being accompanied with difficulty of breathing and palpitation of the heart. It is usually caused by some organic difficulty of the heart, or its arteries, although, at times, it is altogether sympathetic, or dependent on disordered nervous action, and in the latter case admits of a cure. When the vessels of the heart are ossified, its substance enlarged, or its cavities dilated, medicine is rather palliative than curative. The paroxysms frequently come on after physical or mental exertion, more especially if such exertion is made after a full meal; and they are usually observed to occur with those who are inclined to obesity, and who possess that bodily formation which favors apoplectic attacks. As the disease progresses, the symptoms become more violent and distressing, are induced by the slightest movement, and sometimes without perceptible cause.

The precautionary measures to be adopted by those liable to attacks of this disease are, the avoidance of all mental and physical stimulation, restriction to a vegetable diet, moderate exercise, and early rising. As the seat of the affection may not be easily ascertained, and as the peculiar symptoms demand the supervision of a discriminating physician, little if any permanent advantage is to be expected from domestic treatment.

The chief remedies are the same as in the preceding section, and the administration of doses are also the same. A reference to the article "Asthma"* may, if the above-mentioned medicines fail, lead to the selection of a remedy for the alleviation of some symptoms of this painful disorder.

* See page 98.

CHAPTER III.

DISEASES OF THE BRAIN AND NERVOUS SYSTEM.

1. Headache.
2. Vertigo.
3. Inflammation of the Brain.
4. Apoplexy.
5. Dropsy of the Brain.
6. Epilepsy.
7. Paralysis.
8. Convulsions.
9. Chorea.
10. Neuralgia.
11. Hysteria.
12. Delirium Tremens.
13. Tetanus.

THE organ through which all mental operations are conducted, and from which the nervous energy is chiefly derived, lies beneath and surrounded by the skull, an oval mass of pulpy matter, closely invested by three thin membranous coverings, and throwing off nerves to the extreme organs of sense. The brain consists of three principal divisions, the cerebrum, the cerebellum, and the medulla oblongata. The first forms the largest and uppermost part; the second lies below and behind; the third is level with the second, and in front of it, uniting the brain with the spinal marrow, an elongated cord of the same substance with the brain, that extends the whole length of the back, through a canal formed by the vertebral bones. From

the brain, and its continuation, the spinal cord, proceed numerous white filaments or nerves, that are
distributed to every portion of the frame, and serve as
instruments of communication between the mind, the
body, and surrounding objects. The nerves of sight
smell, hearing, and taste arise from the brain; the
nerves of sensation and motion, with a few exceptions,
from the spinal marrow. They issue in pairs through
the whole distance, one passing to either side; nine
from the brain, which are chiefly appropriated to the
four senses above named; and thirty-one from the
spinal cord, at each junction of the vertebral bones;
and they extend to the surface of the body, to produce
the fifth general sense of feeling. In addition to this
grand centre and source of nervous influence, there is
a collection of nerves, or one great nerve, called the
sympathetic, that arises from the branches of the cerebral and spinal nerves, and passes down on each side
of the spine. This is not strictly referable to the grand
centre, but forms in itself a system, having several
centres or ganglions situated throughout its course,
from all of which radiate fine nerves to the different
internal organs and large blood vessels; thus connecting each to the other by nervous communications. To
this medium of sympathy is due the disturbance that
the stomach experiences when the brain is affected
and the reverse; as also the sensitive fellowship of
feeling exhibited between other organs, remote from
as well as contiguous to each other. The action of
the nerves is reciprocal; first, from without inwards,

as external impressions on the organ of sense are communicated by the nerves to the brain, and then give
rise to perceptions; second, from within outwards, as
all voluntary motion is produced by communication
from the brain to the nerves that are freely distributed
over the muscles of the body. All physical action
depends on the nervous energy transmitted from the
brain through the nerves, and when the latter are
diseased, communication is interrupted, and the sense
to which they are specially appropriated is diminished
in capacity, or altogether destroyed. · In spasmodic
affections, the control of the mind over the nerves of
motion is lost.

There are many difficulties attending the investigation of nervous diseases, as the causes are not often
easily ascertained, and the conditions are variable and
inconstant. Those affections from which the most severe pain is experienced are frequently unattended by
any traceable deviation from the natural condition of
the nerves. Even life may be destroyed by an influence exerted on the nervous system, while no vestige
of local, structural injury is to be perceived. There
are few diseases of any sort which do not more or less
derange nervous action, and it is extremely difficult,
in very many cases, to discover to what extent the
derangement proceeds, how much is due to sympathy,
and whether the disturbance is primary or secondary.
It is undeniable that nervous diseases are more prevalent now than formerly; that the nervous system is
more excitable, more disposed to local and general

derangement, in civilized countries. Whatever other causes may have operated to the production of this susceptibility, it is very plain that a departure from the simple and natural habits of a previous age forms one, if not the principal one. The modern method of cookery, the stimulating articles introduced into food, the excessive use of coffee, tea, and alcoholic drinks, to say nothing of the mad struggle for wealth, have contributed much towards the prevalence of nervous maladies. Constant stimulation, of whatever nature, deranges the brain, disturbs the whole process of digestion, disorders the secretions, accelerates the circulation, causes congestion, gives rise to hysterical and hypochondriacal affections, epilepsy, apoplexy, and paralysis. Such consequences, however, embarrassing as they generally prove to every practitioner, more especially of the old school, are more controllable by homœopathic treatment than by any other method; the details of which, in relation to individual affections, are now to be given.

HEADACHE. (*Cephalalgia.*)

This complaint is usually a symptom of some other affection, more especially of a disordered stomach, though sometimes idiopathic, or independent of disease in any other location. It is not always easy to determine whether it is symptomatic or idiopathic. There are numerous remedies, each applicable to the various forms of headache, to be selected according

to the presumed cause, and the character of the pain
That form which is evidently symptomatic of some
derangement of digestion is the most general, while
there are others of a rheumatic, nervous, and congestive origin.

It will be necessary to refer separately to the presumed causes, and to enumerate the medicines most
appropriate to their several consequences. As has
been stated, the most common is

HEADACHE FROM INDIGESTION.

This is usually accompanied with loss of appetite,
unpleasant taste in the mouth, lassitude, furred
tongue, a sense of weight or oppression in the stomach, nausea or vomiting, the pain being generally a
deep-seated, dull aching over the brow.

The remedies are, —

1. NUX VOMICA.

Pain in forehead, sometimes over left eyebrow; increased by eating, by moving, by stooping, by noise
and heat; growing worse towards mid-day, and relieved by lying down.

2. PULSATILLA.

Pain similar to the above, but worse in the evening;
increased by rest, by moving the eyes; relieved by
walking, and by pressure upon the head.

3. IPECACUANHA.

If the above pain is attended by constant nausea, or
vomiting.

DOSE. — Six globules may be taken dry, at intervals of two, four, six, or eight hours, according to the severity of suffering, and the medicinal impression produced.

HEADACHE FROM A COLD.

Pain in the head is sometimes a principal catarrhal symptom, and is attended with a degree of febrile action. When this is the case, the medicines to be taken are, —

1. ACONITE.

Heat alternating with chills ; violent pain, increased by the touch, and by talking ; irritability.

2. MERCURIUS.

Pressing pain in forehead, with symptoms of a catarrh that is epidemic ; pain in the limbs ; tendency to perspire at night, with thirst.

3. NUX VOMICA.

Heavy pain in forehead, with nasal obstruction, particularly at evening; heat in head and body; a sensation of soreness of the skin ; constipation.

DOSE. — Same as for the preceding.

HEADACHE FROM CONGESTION.

Pain in the head, in consequence of an unusual fulness of the blood vessels, or a determination of blood to the brain, is distinguished by a throbbing, pulsative sensation in forehead and temples, dizziness, humming

in the ears, and heat in head, with flushed face, confu
sion of thought, inability to stoop without great aggravation of the sense of fulness, and of the throbbing
pain. When such symptoms are present, the remedies
are, —

1. BELLADONNA; 2. ACONITE;

When the above symptoms are attended by a full,
rapid pulse, with general febrile action.

When there exists a constant tendency to this disturbance, the medicines named below may be given,
one dose every third day, viz. : —

1. CALCAREA; 2. BELLADONNA;

In alternation.

DOSE. — Same as for the preceding.

RHEUMATIC HEADACHE.

For fugitive rheumatic pains in the head, aggravated
by movement, worse at night, attended with great sensitiveness to touch, perspiration, and sometimes vomiting, the remedy is, —

1. NUX VOMICA.

Taken every fourth hour; to be followed, if ineffectual after twenty-four hours, by, —

2. CHAMOMILLA; 3. PULSATILLA;

In alternation.

DOSE. — Same as for the preceding.

NERVOUS HEADACHE.

In this species of headache there are no signs of inflammation, but the face is pale, the head cool, the pain is darting, pricking, or crawling, frequently confined to one side of the head, with a sensation as if a nail were driven into the brain. For such symptoms, the remedy is, —

1. COFFEA.

Especially if attended with nervous symptoms, restlessness, sensitiveness to light and noise, irritability, anguish, weeping.

2 CHAMOMILLA.

For above symptoms when not relieved by *Coffea*, in consequence of the daily use of coffee.

3. PULSATILLA.

In addition to the above, darting pain in the ears, thirst, dizziness, partial blindness, heaviness of the head, and palpitation of the heart.

DOSE. — Same as for the preceding.

SICK HEADACHE.

The kind of headache so named is attended with continued nausea, and not unfrequently terminates, if left to itself, in vomiting. It is of a periodical character, occurring at intervals of a month, more or less, and is most common to persons of delicate health and irritable temperament. The pain is generally confined to the

10*

forehead, or one side of the head. It is not always to
be controlled, but generally great relief may be had by
the use of, —

1. PULSATILLA.

Especially if the pain is increased by lying down, and
relieved by walking; want of appetite; sadness; palpitation ; dizziness.

2. VERATRUM.

Chilliness ; painful sensibility of head to touch ;
great debility ; deep-seated internal heat in head ;
nausea and vomiting.

Dose. — Of all the above medicines for headache, six
globules may be taken, dry, at intervals of two, four,
six, or eight hours, according to the severity of suffering, and the medicinal impression produced.

CHRONIC HEADACHE

Is best treated by

1. CALCAREA; 2. SULPHUR; 3. SEPIA.

Dose. — Six globules every evening for three days,
then omitting, to be resumed after an interval of the
same period.

Diet. — Coffee and tea should, in all periodical headaches, or such as occur frequently and for a long time,
be given up entirely. Articles of food that have ever
been found to disagree with the stomach, must not be
used. All spices and oils should be avoided, and daily
exercise taken in the open air.

VERTIGO. (*Dizziness.*)

Though this sensation is nothing more than symptomatic, to be removed only by acting against its cause, yet, when the cause is not ascertained, the following medicines will moderate the dizziness, more especially if occurring on rising from a recumbent position, or when stooping.

1. ACONITE; 2. BELLADONNA;

In alternation.

DOSE. — Six globules dissolved in ten spoonfuls of water, and one spoonful given every three hours.

The same remedies may be given in the same man ner for

DETERMINATION OF BLOOD TO THE HEAD.

Indicated by fulness of the vessels of the head and neck. heat and redness of face, attacks of giddiness after sleeping or exposure to warmth of the sun or fire, dimness of sight, buzzing in the ears, drowsiness, and headache.

DIET. — The food should be light and easy of digestion with those in whom these disturbances prevail. Early rising and daily exercise will prove beneficial.

116 HOMŒOPATHY SIMPLIFIED.

INFLAMMATION OF THE BRAIN. BRAIN FEVER.
(*Phrenitis.*)

The brain itself, as well as its investing membranes, is subject to inflammation, which, when once established, is of a serious character. It is therefore impor-tant to anticipate its attack, and to mitigate the inflam-matory action in its incipient stage; the symptoms attending which are heat, pain, and heaviness in the head, flushed, swollen face, eyes suffused and brilliant, thirst, throbbing of the vessels of the neck, irritability, contracted pupils, giddiness, restlessness, intolerance of light, nausea, and sometimes vomiting; in short, symptoms of determination of blood to the head. For this condition, the first medicines to be given and con-tinued until improvement or a decided increase of the disease takes place, are, —

1. ACONITE; 2. BELLADONNA;
In alternation.

At a later stage, the following are to be used in ac-cordance with the indications : —

3. HYOSCYAMUS.
Loss of consciousness; delirium; dilated pupil; in-articulate speech; parched skin; convulsive move ments; great drowsiness; picking of the bed clothes.

4. STRAMONIUM.
Sudden starting as from fright; quiet sleep, with no cognizance on awaking; fixedness of look; redness of face; high fever, with a moist skin.

Reference : —

OPIUM; CUPRUM.

In inflammation of the membranes, the pain is more violent than when the brain only is inflamed. There are but few reliable -marks of distinction — a circumstance entirely unimportant, as the treatment is the same in both cases.

DOSE. — Six globules to be dissolved in six spoonfuls of water, and one spoonful given every hour, and at longer intervals, in proportion to the degree of improvement.

DIET. — In this dangerous disease the diet should be of the most simple kind — exclusively liquid, until inflammation is over. Cold water is the best drink, and applications of cold water may be constantly made to the head, which should be raised higher than usual when lying down.

APOPLEXY.

The immediate cause of this serious affection is a compression of the brain, resulting either from an inordinate accumulation of blood in the vessels of this organ, or an effusion of fluid through the coats of these vessels. This condition may be produced by any action which determines blood to, or interrupts its reflux from, the head, such as violent muscular exertion, long-continued stooping, powerful mental emotion, compression of the veins of the neck. The characteristic symptoms are an immediate abolition of sense and

motion, snoring respiration, swollen face, strong, full
pulse, distended blood vessels of the face and neck,
and dilated pupils.

In all its varieties this disease is alarming; and if
occurring in aged persons, seldom terminates favorably.

There is much reason to believe that the practice
of bleeding, which at one time was considered an
indispensable expedient, has, instead of restoring consciousness, frequently destroyed it altogether. Drunk-
ards are subject to apoplexy. It has also been induced
by the inhalation of metallic or narcotic substances,
and by external injuries. The attack is often preceded by the following symptoms: Drowsiness, fatigue
after slight exertion, vertigo or fainting, pulsation of
the arteries of the neck, dimness of vision, indistinct-
ness of speech, forgetfulness, irritability, numbness or
prickling sensation in the limbs, headache. For these
symptoms of congestion, as well as for the attack itself,
the most appropriate medicines are, —

1. ACONITE; 2. BELLADONNA;

In alternation.

After which, if required, —

3 NUX VOMICA; 4. OPIUM;

In alternation.

DOSE. — For the premonitory symptoms, twelve
globules to be dissolved in a wine-glassful of water,
and a spoonful given every four hours alternately;

while for the immediate attack, a spoonful of the same is to be given every fifteen minutes.

DIET. — To obviate the predisposition to this danger-ous disease, hearty eating should be avoided and stim-ulants altogether dispensed with. During an attack, the head should be raised, the dress should be loos-ened about the neck and chest, pure, fresh air freely admitted, and the feet placed in warm water, or rubbed hard with warm cloths.

DROPSY OF THE BRAIN. (*Hydrocephalus.*)

This affection is almost exclusively confined to chil-dren, rarely occurring after the age of fourteen, and generally attacks such as are of a scrofulous habit. Its symptoms are, pain in the head, slow pulse, nausea and vomiting, stupidity, dilated pupils, distortion of eyes, with convulsions; in many respects resembling those of inflammation of the brain. In some in-stances the disease comes on slowly and gradually, commencing with a slight indisposition, during which peevishness and a disinclination to be raised from bed are quite noticeable, and terminating in the symptoms above mentioned. When gradual in its progress, the head becomes much enlarged and altered in shape, in consequence of the accumulation of water. The dura-tion of the disease varies from two to six weeks. When extending beyond the latter period it must be considered as chronic, and in this form may continue for years. The remedies most appropriate for the pre-cursory symptoms are,—

1. ACONITE; 2. BRYONIA;

In alternation.

When the disease is established, the medicines the most frequently adapted are, —

1. HELLEBORUS; 2. SULPHUR.

DOSE. — The former, viz., *Aconite* and *Bryonia*, may be administered morning and evening, three globules of each, dry; the latter in the same dose every third hour.

DIET. — The diet should be simple and easy of digestion. The head should be kept cool and the feet warm.

EPILEPSY.

The attacks of this disease are sudden, and manifested by convulsions, distortion of the eyes, clinching of the fingers, unconsciousness, and almost always by foaming at the mouth. The causes are various; as, external injuries of the head, internal tumors or other morbid growths pressing on the brain, sudden and violent emotions and passions, teething, worms, poisons, &c. In some cases the paroxysms come on without manifest cause. Preceding an attack there is often experienced a peculiar sensation, like that of cold air gradually ascending from the lower limbs to the brain. Should epilepsy be symptomatic of any know

disturbance, like the presence of worms in the ali
mentary canal, deranged digestion, dentition, poison,
&c., the exciting cause must be acted upon by the
appropriate remedies; and in such cases the attacks
are easily subdued. But when from a constitutional
disturbance, or when of frequent occurrence and long
duration, or occurring late in life, it will be difficult
to effect a cure.

The medicines chiefly indicated are,—

1. ACONITE; 2. BELLADONNA;

In alternation, for premonitory symptoms.

After the attack, the principal medicines to prevent
the recurrence, to counteract the predisposition to the
complaint, are,—

1. CALCAREA; 2. SULPHUR.

During the paroxysm, the person affected must be
placed in a recumbent position, all clothing about
the neck and chest loosened, cold water sprinkled
over the face, and every object removed that might
cause injury during the convulsive movements of the
head and limbs.

DOSE.—Of the first, eight globules dissolved in a
wine-glassful of water, and a spoonful given every
three hours; of the latter, six globules, dry, morning
and evening.

DIET.—Persons liable to these attacks should live
on simple, digestible food, avoiding stimulants and
overloading of the stomach, and violent exertion,
either physical or mental.

11

PARALYSIS. (*Palsy.*)

This may arise from any cause that obstructs the
communication from the brain to other parts. It is a
loss of motion and sensation, complete or partial,
affecting either the whole or portions of the body;
and the attack is sometimes immediate, but more
generally preceded by a sense of coldness and numbness, as also occasionally by spasmodic action. This
loss of sensation and motion, or both, will be proportionate to the influence of the morbific agent. When
resulting from apoplexy, when any vital organ is attacked, or when caused by an injury done to the
spinal marrow, in which case the parts below the seat
of the injury are paralyzed, the disease is seldom
within the control of medicine. But in many instances it is curable, especially when organic life is
unimpaired; when owing to debility, or a transfer of
morbid influence, as scrofula, rheumatism, &c., to the
nerves; and when the complaint has not been of long
duration. It is well known that a paralytic state is
induced by the absorption of *lead* into the system, as
they who by their occupation are exposed to the
poisonous action of this mineral suffer more frequently
than others from palsy. The medicines that have
proved the most useful in this disease are, —

1. ARNICA.

When one side is affected; weaknes of 'oints, with

swelling; great debility, with sense of fatigue ; when succeeding a rheumatic attack.

2. BRYONIA.

Limbs paralyzed, with oppression of the chest and laborious breathing; loss of motion, but not wholly of sensation; a sensation of cold in limbs.

3. SULPHUR.

Hip joint paralyzed, with pain extending to the back, attended by itching eruption on the skin ; palsy following a repelled eruption.

For poisoning by lead, —

OPIUM; BELLADONNA.

DOSE. — Six globules of the medicine selected, taken in a dry state, morning and evening.

CONVULSIONS.

This morbid condition, consisting of an alternate contraction and relaxation of muscular fibre, is dependent on the state of the brain and nerves, and is especially common in childhood, when the nervous system predominates. Irritation from dentition, worms, mechanical injuries, mental emotion, as anger, fear, &c., are among the exciting causes. "Spasm" is of a similar origin and nature, but differs in being muscular contraction where there is no tendency to alternate relaxation. Convulsions may be general **or** partial, receiving names according to the parts affected. The principal remedies for convulsions are, —

1. CHAMOMILLA; 2. BELLADONNA;

In alternation.

In case of failure of the above, the next best medicines are, —

3. IGNATIA.

Sudden flushes of heat, with general trembling; violent crying; when the fit returns at regular hours.

4. ARSENICUM.

Burning heat; insatiable thirst; irritability; diarrhœa, or vomiting.

References: —

NUX VOMICA; PULSATILLA; SULPHUR.

DOSE. — Three or four globules should be placed, dry, upon the tongue, and repeated every fifteen minutes during the paroxysm.

The child should also, in addition to the frequent administration of the above, have the lower half of the body immersed in hot water for ten minutes, then wiped dry, and wrapped in warm flannel; and if the fit continues, a small stream of cold water should be allowed to fall on the top of the head and back of the neck for twenty or thirty seconds, — to be renewed at intervals of five minutes thereafter, for the space of an hour. When animation returns the child should be warmly covered. If there is reason to suspect undigested food as the cause of the convulsions, warm water injections should be freely administered. Particular attention should be given to the quantity and quality of food when there is a tendency to convulsions.

CHOREA. (*St. Vitus's Dance.*)

This is a species of convulsions principally confined to the limbs of one side, and is caused by an irritability of the nervous system from indigestion, intestinal irritation, mental disturbance, repelled eruptions, and sometimes from the force of sympathy. It generally attacks children, and habits debilitated by medicine or other causes at the age of puberty. At any period of life, however, when there is extreme nervous irritability, an irregularity of nervous action, involuntary convulsive movements may occur, either partial or general. The muscles of the face are often the first to be attacked, and the limbs are afterwards affected. Chorea is not accompanied with danger to life, and is generally curable; but local convulsions, especially of the face, sometimes become permanent, resisting every mode of treatment. The medicines that have proved most successful in the removal of this troublesome affection are,—

1. COFFEA; 2. BELLADONNA;

In alternation.

DOSE. — Three globules, dry, every morning and

11 *

evening, — the first in the morning, the second in the evening.

The trial of *Electro-Magnetism* has been recommended in obstinate cases.

NEURALGIA.

This term is used to designate a disease confined to certain branches of the nerves, as those of the face, chest, and limbs. Neuralgia of the nerves distributed to the face has been termed " tic douloureux." The nerves passing to all portions of the body are subject to attacks of this painful disorder; and wherever it may be located, there are but few complaints to which the human race is exposed that are attended with such intolerable suffering, or more obstinate to medicinal action. The pain is excruciating, shooting along the course of a nerve or branches of nerves, attended with more or less inflammation about the affected nerve, and sometimes causing an involuntary contraction of the adjacent muscles. The chief remedies are, —

1. ACONITE.

Pain most severe at night; feverishness; thirst; redness of cheeks; sight and hearing painfully sensitive; pain in the region of the heart, with great anguish.

2. BELLADONNA.

Shooting pain, increased by warmth, air, and noise; redness, heat, and swelling round the affected nerve;

trembling of the limbs; headache; pain in the back; for plethoric habits.

3. COFFEA.

Extreme sensitiveness to the slightest noise; sadness or ill-humor; sleeplessness; dread of cold air; great agitation.

4. PULSATILLA.

Pain pulsative, felt on one side, increased by rest, lessened by the external air; pain in the region of the heart or stomach; adapted to women and those of a mild, timid disposition.

References : —

MERCURIUS; NUX VOMICA; VERATRUM.

DOSE. — One spoonful of a solution containing six globules in twelve spoonfuls of water may be given every hour until relief is obtained.

The *Tincture of Aconite* has been strongly recommended, three drops in a wine glass of water, a tea spoonful to be given every fifteen minutes, together with the external application of the *Aconite* to the part affected.

That pain which is consequent on a wound of the nerves may be relieved by *Arnica* or *Calendula;* the former when the injury is caused by violent contact with a blunt object, and the latter when produced by a cutting instrument.

HYSTERIA. (*Hysterics.*)

This singular affection, dependent on a morbidly increased sensibility of the nervous system, and manifesting itself by a great variety of symptoms, is peculiar to the female. An attack commonly commences with difficulty of breathing, sadness, weeping, or laughing without cause, and in rapid alternation, accompanied with extraordinary and often most ludicrous fancies.

A very frequent cause of complaint is the " globus hystericus," a sensation as though a round ball ascended from the region of the stomach and lodged in the throat, causing a dread of suffocation ; soon the limbs are spasmodically affected, and the body distorted into every variety of position ; or, as in some instances, partial or total insensibility comes on, and the affected person lies for hours perfectly motionless. Great is the variety of forms, however, assumed by this nervous affection. The leading remedies are, —

1. COFFEA.

When attacks occur during menstruation, with great agitation, cold perspiration, spasmodic movements, and screaming.

2. IGNATIA.

Extreme sensitiveness to light and noise ; hard, swollen abdomen ; paleness of face ; nausea ; chilliness, and drawing pains.

3. PULSATILLA.

When during menstruation, an aggravation of symp

toms is felt towards night; in a mild, gentle nature, inclined to sadness.

References: —

BELLADONNA; HYOSCYAMUS; NUX VOMICA.

Dose. — During the paroxysm, a spoonful of water in which one globule is dissolved should be given every fifteen minutes; afterwards every four to six hours, to prevent their recurrence.

Diet. — Coffee, tea, and all stimulants to be avoided. In the attacks the clothing should be loosened about the throat and waist, pure air freely admitted, and cold water sprinkled upon the face.

DELIRIUM TREMENS.

The prominent symptom of this affection, chiefly consequent on the intemperate use of alcoholic drinks, is sleeplessness accompanied with some degree of mental derangement. The mind is constantly disturbed by frightful visions, and fixed ideas of approaching evil. The speech is inarticulate but incessant; the limbs tremulous, and occasionally convulsed; the expression of the face is wild and variable, with rolling, restless eyes, and the skin is covered with perspiration. These symptoms are usually preceded by great irritability, loss of memory, anxiety, and extreme restlessness of mind and body. This condition almost invariably results from the deprivation of a powerful stimulus, to which the brain and nervous system have been daily and for a long time accustomed. The known previous habits of the person affected will sufficiently indicate the nature of the delirium, which otherwise might possibly be confounded with that accompanying inflammation of the brain, which, in some respects, it resembles. The medicine best adapted for its treatment during the first stages is

NUX VOMICA.

When the delirium is fully established,

OPIUM.

The above, after having .been administered singly for two days successively, every two hours, may afterwards be given in alternation every four hours; and, in ordinary cases, nothing more is required to remove the disease.

DOSE. — Twelve globules dissolved in six spoonfuls of water, one spoonful of which may be given for a dose.

The best means to counteract an inordinate inclination for alcoholic stimulants is believed to consist in the frequent use of *Sulphuric Acid*, mixed with water, in such a proportion as will make the drink slightly acid, and given freely in every thing that is taken until the mouth becomes sore, and complete derangement of digestion is produced.

TETANUS

This is a general spasm of the whole body, characterized by strong, continued muscular rigidity. The head is bent forward or backward so far and forcibly as sometimes to touch the feet. When the rigidity is partial, and confined to the muscles connected with the mouth, closing the jaws so firmly as not to admit of the smallest opening, it is termed " trismus," or " locked jaw." Its attacks are sometimes sudden and violent. But they usually come on in a gradual manner, with a stiffness at the back of the neck, which renders the motion of the head difficult and painful, the stiffness by degrees extending to the jaws. If proceeding beyond this, the muscles of the whole body soon become forcibly contracted, and to this condition the term " tetanus " is applied. This painful disorder is occasioned either by exposure to cold, or by some local injury to the nerves, from amputation, puncture, laceration, &c. The chief remedy for this affection, whether general or partial, is, —

ARNICA,

Applied externally, near the seat of the iniury, a table-spoonful of the tincture in a tumbler full of water; the same medicine to be given internally as below indicated. *

Reference : —

BELLADONNA; PHOSPHORUS; SULPHUR.

DOSE. — Ten globules in eight spoonfuls of water, and one spoonful given every hour.

* For making *Tincture of Arnica,* see directions at the end of the book.

CHAPTER IV.

ON FEVERS.

1. Simple Fever.
2. Inflammatory Fever.
3. Typhus Fever.
4. Bilious or Gastric Fever.
5. Intermittent Fever.

ERUPTIVE FEVERS.

6. Small Pox.
7. Varioloid.

8. Chicken **Pox**
9. Measles.
10. Scarlet **Fever.**
11. Scarlet **Rash.**
12. Miliary **Fever.**
13. Nettle **Rash.**
14. Erysipelas.

THE human system possesses an inherent capacity of resisting whatever tends to disturb its natural condition ; and through the action of this power of resistance in attempting the expulsion from the body of disease-producing influences, a general commotion is caused which is termed " fever." If any particular organ or set of organs are more especially attacked, the resisting energy will be there concentrated, and the part seized upon will suffer in proportion to the strength of the invading force. The locality of the contest often determines the kind of fever. If the struggle is going on in the brain, the lungs, the liver, &c., the fever or inflammation obtains its name from the organ affected, as brain fever, lung fever, inflammation of the liver, &c.

Homœopathic remedies coöperate with the restorative efforts of Nature, and overpower disease before any great amount of vital energy has been expended. The assistance they render is most prompt and efficient, never disturbing or destructive, but ever quiet and gentle in their operation. By such aid, Nature rapidly regains her normal condition, the reëstablish ment of health being thorough and permanent.

The external signs by which fever is manifested are, preternatural heat and redness of the skin, frequent pulse, coated tongue, thirst, restlessness, languor, and deficient secretions. There are various kinds of fever, various phases of the same fever, and various influences that give rise to them. The causes in general are exposure to atmospheric changes, unwholesome diet, vitiated air, intemperance in eating and drinking, over-exertion of body or mind, and contagions of different kinds.

DIET IN FEVERS. — During the whole course of a fever, mental and bodily rest must be preserved, the patient be kept in a moderate temperature, in pure air, and under bed clothing not too warm or heavy. The diet should be strictly simple. No animal food must be taken; nothing that can stimulate or excite the digestive organs to action. The want of appetite during acute diseases is an indication that the stomach is not in a proper condition to receive food, as the prevailing thirst is equally an indication that the .

12

system requires water. The return to solid food during convalescence should be gradual, as vigorous digestion is not likely to be reëstablished until the restoration to health is complete. The erroneous idea that some substantial nutriment is necessary for the support of life during the continuance of acute febrile . action has, when practised upon, resulted in consequences, oftentimes, of a very serious nature; and protracted recoveries, with relapses into increased debility, may, not unfrequently, be with justice attributed to the hasty resumption of solid diet. When on the subsidence of fever the appetite returns, light fluid nutriment, such as barley water, gruel, porridge, and the like, should at first be given in small quantities; and afterwards arrowroot, sago, rice, broth, and articles of such a nature, until the digestive organs are enabled to act in a healthy manner, when the customary diet may be resumed. In general, the wishes of the patient should be consulted; and no kind of food should be forced upon him which may be unpleasant, or which, in the usual state of health, has been found difficult of digestion.

SIMPLE FEVER.

This species is the lightest of all fevers, and of the shortest duration, seldom existing, when distinct, longer than a day. Its symptoms are chills, suc ceeded by heat, thirst, quick pulse, restlessness, with. debility; and the fever terminates by copious perspiration. This group of symptoms, whenever occurring, either independent or introductory to any continued fever, indicates invariably, and requires no other remedy than

ACONITE.

DOSE. — Twelve globules dissolved in a tumbler half full of water, a spoonful of which is to be given every two, four, or six hours, according to the degree of fever present.

DIET. — There will be no appetite during the short continuance of the simple fever, and nothing, of course, need be given more than cold water, which may be drank freely.

INFLAMMATORY FEVER.

This form of fever attacks with more severity and suddenness than the preceding. The commencing chill is considerable, the heat of surface extreme, the pulse strong, hard, and very rapid, the thirst great, the respiration hurried, the face and eyes red, the skin dry, the tongue coated white, and the secretions obstructed.

Persons of a full, plethoric habit are the most liable to this species of fever. Its chief causes are, obstructed perspiration from cold or dampness, the free use of stimulating food, strong mental emotions, repelled eruptions. Its duration is from two to fourteen days, and if not moderated by medicines at its commencement, is attended during its progress by some degree of delirium. The principal, and, in a large majority of cases, the only medicines needed are, —

1. ACONITE; 2. BELLADONNA;

In alternation, every two hours.

Should the lungs and stomach be more affected than the head, as indicated by very difficult breathing, pains in the chest, or nausea, and obstinate costiveness, the following will be more applicable, viz. : —

1. ACONITE; 2. BRYONIA;

As above.

DOSE. — A spoonful every two or three hours of a solution made by dissolving eight or ten globules in four table-spoonfuls of water; a tea-spoonful of *Aconite* to be first given, two or three hours after which, a spoonful of *Belladonna;* so continuing until the severity of the fever abates, when the interval is to be lengthened accordingly.

DIET. —. Nothing but cold water is needed, until the restoration of appetite and abatement of fever. A return to solid food should be preceded by gruels, arrowroot, &c., and afterwards the different kinds of broths When the skin is hot and dry, sponging the body with cold water will be serviceable and grateful.

TYPHUS FEVER.

To the various grades and phases of this continued fever have been applied the terms "typhoid," "nervous,"

" slow," " putrid," " ship," "jail," " camp," and " hos-
pital " fever. In its mildest form, it is known as the
" typhoid," " slow," or " nervous " fever. The remaining terms are applied to it when malignant, while the
general and true appellation is " typhus." It is a fever
of debility, with a tendency of the fluids to putrefaction ; and it is distinguished from other fevers by the
weakness of pulse and great prostration of strength at
its commencement and during its progress. The symptoms are pains in the head, back, and limbs, heat and
dryness of the skin, extreme mental and bodily depression, thirst, constipation, delirium, all more or less
severe, in proportion to the violence of the fever. Its
duration, in its various forms, is from two to six weeks,
and even longer. The treatment of the malignant
typhus should be resigned, of course, to the management, if accessible, of the educated physician ; but in
its incipient stage, it may be greatly mitigated, if not
wholly arrested, and throughout its course, in the mild
form of slow fever, it may be controlled by the use of, —

1. BRYONIA; 2. RHUS TOXICODENDRON ;

In alternation.

Preceded in all cases, when inflammatory symptoms
present themselves, by

1. ACONITE; 2. BELLADONNA;

In alternation.

In its severer forms, the chief medicines to be consulted are, —

1. ARSENICUM; 2. CARBO VEGETABILIS; 3. SULPHUR.

13 *

Dose. — Ten globules dissolved in four table-spoon-fuls of water, a tea-spoonful of which is to be given alternately every three hours.

Diet. -— Pure cold water is all that is at first required. As the thirst subsides, and fever abates, the more nutri-tious fluids, as toast water, gruel, arrowroot, &c., may be given. When the appetite for more solid food re-turns, caution should be exercised in its gratification. Sponging with cold water may be employed at all times when the skin is hot and dry. Stimulants, as wine-whey, or brandy and water, may be allowed in small quantities and at short intervals, when the skin becomes cold and clammy, the pulse intermits, and the breath-ing hurried and oppressed. The sick room should be large and well ventilated, the air preserved as pure as possible.

BILIOUS FEVER. (*Gastric Fever.*)

In this form of fever, the organs of digestion are in-volved, especially the liver. It differs in character from those hitherto described, being remittent, that is, having intervals during which febrile action is partially suspended, the patient for some hours, chiefly in the morning, remaining comparatively free from fever. The principal symptom of the bilious or gastric fever is, great irritability of the stomach, inducing vomiting, sometimes of a very obstinate kind, provoked by the presence of the smallest quantity of food in the stomach. This is attended either with constipation or diarrhœa, and a yellowish hue of the skin, with alternate chills and flushes of heat, and but slight, if any, perspiration. There is both a mild and severe bilious remittent, as well as a mild and severe form of the preceding disease (typhus); a general disturbance of the nervous system distinguishing the former, and manifestations of putrescency and extreme exhaustion the latter. At first, this distinction does not always plainly prevail, but after a time the nervous or putrid characteristics become very evident. The principal medicines in the mild form of this fever, are, —

1. CHAMOMILLA; 2. NUX VOMICA;

Given in alternation.

In the severe forms, the remedies most often indicated are, —

1. CARBO VEGETABILIS; 2. ANTIMONIUM TARTARIUM;

Also in alternation.

Reference : —

ARSENICUM; VERATRUM.

DOSE. — Six globules in a wine glass of water, and one spoonful given every two or three hours.

DIET. — The food must be light, chiefly fluid; in fact, nothing is desired more, while the bilious symptoms predominate, than water; and even that cannot always be retained.

INTERMITTENT FEVER. (*Fever and Ague.*)

The term "intermittent," as applied to fevers, signifies a cessation of febrile symptoms for a certain interval, and differs from the above described remittent form in the cessation being complete instead of partial. This peculiar kind of fever generally occurs in newly-settled districts of country, where the earth, that has remained long undisturbed, is exposed to the sun and air for the purpose of cultivation.

This form is manifested by very distinct periods of chilliness, heat, and perspiration ; and paroxysms including these three stages take place daily, regularly returning every twenty-four hours — then called quotidian ; or every other day, after an interval of forty-eight hours — then called tertian ; or after an interval of seventy-two hours — then called quartan ague. These divisions are also subject to variations and subdivisions, which, however, in a · practical sense, are unimportant. The desirable object is, to discover the duration of the intermission, in order to anticipate the return of a paroxysm, since the remedies are to be employed during this intermission.

During the chill, or first period, external warmth may be applied in every convenient way, and warm, diluent drinks given. During the heat, or second period, applications of cold water to the surface, and internally, as well as a few doses of *Aconite*, when the fever is violent. As soon as the sweating, or third period, is over, the medicine most appropriate to the characteristics of the disease must be given and continued at regular intervals while the intermission lasts.

Such medicines are, chiefly, —   ·

1. **CINCHONA.**

If from direct exposure to marsh miasm ; and if there is headache, nausea, great hunger or thirst, yellowness of the face, and pain in the abdomen.

2. **IPECACUANHA.**

Increase of chills by external warmth ; want of thirst ; nausea and vomiting.

3. **ARSENICUM.**

Great thirst ; external burning heat ; burning pain in stomach and limbs ; prostration of strength.

Reference : —

NUX VOMICA; PULSATILLA; ANTIMONIUM; TART.

DOSE. — Twelve globules dissolved in six spoonfuls of water, and one spoonful given every three hours.

DIET. — Wheat meal bread, dry toast, a moderate quantity of vegetables, and plenty of fruit, may be used, with but very little animal food. All kinds of stimulants should be avoided. Cold water may be drank freely, and moderate exercise taken.

ERUPTIVE FEVERS.

By this term is meant that class of disorders which commence with the ordinary febrile symptoms, and soon manifest their peculiar character by an eruption on the surface of the body. Such are the small pox, chicken pox, measles, scarlet fever, nettle rash, miliary fever, and erysipelas. Most of these diseases are produced by a specific contagion, and the fever preceding

them is symptomatic. The nature of the contagion
baffles research, and it is perceptible only in its results.
These are the characteristic eruptions, for the development, maturing, and removal of which all vital energy
is directed to the surface. When large, irritating doses
of medicine are administered, the restorative action engaged in removing the infection through the surface is
interfered with, and transferred to the stomach and
bowels, to oppose the violence there committed. Instead of withdrawing Nature's forces, thus weakening
her capacity of resistance by an assault on another
position, homœopathy affords aid by direct coöperation
with the recuperative power, and through this combined action the enemy, disease, is promptly and suc-
cessfully resisted. The superior safety, as well as
energy, of the homœopathic treatment over all other
methods known, has been particularly demonstrated in
the management of eruptive fevers.

SMALL POX. (*Variola.*)

This febrile disease is attended with an eruption of
pustules, which appear on the third day after the inva-
sion of fever, in small red spots on the face, neck, and
breast, that increase in size and number for three or
four days, when they are found dispersed over the
whole surface of the body. These inflamed spots, appearing at first as pimples, enlarge as they proceed to
suppuration, and are distinguished from other eruptions by a depression in their centre. When they

remain separate from each other, the eruption is called "distinct;" when they join each other, it is called "confluent."

An attack of small pox does not at first differ in its symptoms from the ordinary manifestations of a violent fever, such as chills, lassitude, headache, hot and dry skin, rapid pulse, coated tongue, pain in the back and limbs, nausea, sometimes vomiting and stupor. One symptom, however, is present, which does not usually exist in fevers, viz., great pain in the region of the stomach, which pain is aggravated by pressure. In the confluent form the fever is more violent, the debility is extreme, and delirium is common, with often an exhausting diarrhœa, while the pustules cluster together and present a dark appearance, instead of the yellowish hue of a proper suppuration. But the most important difference between the two forms consists in what is termed the " secondary fever," which takes place just when the desiccating stage commences. This fever is scarcely noticeable in the distinct form, but prominent and perilous in most cases of the confluent. During the course of small pox there are four stages: the *febrile*, when fever predominates, lasting three or four days; the *eruptive*, from the first appearance of the eruption until it takes its distinguishing feature, the indented surface, lasting four or five days; the *maturative*, when the pustule suppurates, or becomes filled with pus, lasting about the same time; and the *desiccating*, during which the eruption dries up and falls off, leaving indentations or deep scars on the skin.

Each stage is attended with different symptoms, and demands different medicines.

In the first or febrile stage, the remedies are, —

1. ACONITE;

When general fever exists.   In alternation with

2. BELLADONNA;

When the head is much affected.   Or with

3. SULPHUR;

When, in the course of the disease, the fever runs high, and is not subdued by *Aconite* alone.

During the second and third stages, the following are to be used, according to their respective indications, viz. : —

1. IPECACUANHA; 2. ANTIMONIUM;

In alternation.   For oppression in chest; nausea and vomiting; cough, with accumulation of phlegm in throat.

3. MERCURIUS.

Hoarseness; difficulty in swallowing; diarrhœa, with dysenteric symptoms.

4. BRYONIA.

Derangement of stomach; rheumatic pain in limbs, increased by motion; constipation; irritability.

5. ARSENICUM.

Great thirst, and dryness of mouth; with dark coating on tongue; and extreme debility.

During the period of desiccation, nothing more may

be required than the application of warm water to the skin, and attention to diet.

References : —

CHAMOMILLA; OPIUM; RHUS.

DOSE. — Ten globules, dissolved in four table-spoon-fuls of water, and one tea-spoonful given every two to four hours, according to the severity of the febrile and other symptoms.

DIET AND REGIMEN. — The food should be light and simple, as in all fevers, no solids whatever being allowed during the feverish stage; unless in the mild form, when toasted bread or biscuit, rice, sago, and the like, may be given; but no animal food should be given until health is completely reëstablished. Every thing used as food and drink must be cool; the patient should be lightly covered, and the room be kept cool and dark, the exclusion of light preventing, to a considerable extent, the much-dreaded " pits " or depressions left in the skin by the pustules.

VARIOLOID.

This term is applied to the small pox when it occurs in one who has been subjected to successful vaccination. By this operation the severity of the disease is greatly mitigated, the precursory symptoms being much less severe, and the pustules less regular in their appearance and distribution, disappearing earlier. There is no secondary fever, and the whole train of symptoms is comparatively mild. The same remedies and regimen are required as in the true small pox, the whole treatment, however, being modified in accordance with the milder character of the symptoms.

There is no better illustration of the principle on which homœopathy is founded, than the well-known circumstance that a medicinal influence producing analogous symptoms to those manifested by the small pox, invariably acts as a preventive or modifier of said disease.

The protective or mitigating property of vaccination in relation to small pox, if the operation is properly performed, exists, in most cases, during life; although, under circumstances of peculiar exposure, re-vaccination is an important precautionary measure. The method of performing vaccination is by inserting the end of a quill, on which is a portion of the vaccine virus, under the skin, and allowing it to remain there, for the space of three or four minutes, until absorption takes place; or by moistening the vaccine crust or scab with water, and rubbing the mixture over the skin that has been previously scratched by the point of a knife or lancet. When successful, a minute pimple is observed about the third day after the operation, which gradually enlarges, becomes filled with a clear fluid, presents the peculiar central concavity, and is surrounded by a swollen, inflamed margin. After the tenth day, the cupped or concave surface of the vesicle changes to a more round or convex form, the contained fluid becomes opaque, and the whole gradually dries into a dark crust, which in a few days becomes detached from the skin.

The first vaccination is best performed during the fifth or sixth month of infancy, though if there is danger of exposure to the small pox, it may be attempted earlier. The matter to be transferred from one child to another should be taken from the pustule, if the use of the quill is preferred, about the tenth day, or when maturation is complete. The child from which the vaccine virus is taken should be free from cutaneous eruptions, or any appearance of disease, as experience authorizes the suspicion that diseases have been transmitted from one to another through the medium of vaccination. Whether the circumstance of the transmission of disease in this manner be fully believed or not, there certainly exists sufficient doubt on the subject to justify a precaution that is never difficult to exercise.

CHICKEN POX. (*Varicella.*)

This affection, which is sometimes mistaken for small pox, to which it bears much resemblance, is generally preceded by symptoms of fever, though usually slight, and not often requiring medical treatment. The eruption occurs about the second day after the commencement of indisposition, in small, red pimples, not unlike those of the small pox, but which first appear on the back instead of the face and breast, are sooner matured, more pointed and irregular, and never cupped or indented in the centre. The duration of the eruption is also shorter than that of the small pox, disappearing wholly by the fifth or sixth day, without leaving any mark. Sometimes the fever is considerable, and for its mitigation there must be administered, —

1. ACONITE.

For general fever.

2. BELLADONNA.

If the head is much affected.

3. COFFEA.

For restlessness, sleeplessness, and anxiety.

Dose. — Ten globules, dissolved in four table-spoonfuls of water, and one tea-spoonful given every four hours.

Diet. — Plain, simple food must be given during the existence of the fever.

MEASLES. (*Morbilli*.)

This eruptive disease is infectious, attended by fever, and ushered in by symptoms resembling those of catarrh, as hoarseness, sneezing, cough, difficulty of breathing, thirst, and sometimes nausea. About the fourth day of these symptoms, clusters of small red points, which do not rise above the skin, appear on the face, often in the form of a crescent, soon coalesce, and extend over the whole surface of the body. The accompanying fever does not subside when the eruption makes its appearance, as is the case in general with this class of diseases, but often increases with it. At the expiration of a week, usually, the inflamed appearance changes to a brown hue, and in a few days after, the eruption disappears, the outer skin or cuticle falling off, in the form of minute scales. The disease of measles is one that usually occurs during the period of infancy or childhood, and is generally mild ; although at times it is quite severe, more especially when happening in adult age. The comparative freedom, under homœopathic treatment, from consequences of a serious and permanent character, leads fairly to the inference that such consequences mainly arise from the old system of drugging.

The remedies to be relied on, and which will be sufficient to control the mild form of this disease, are, —

1. ACONITE; 2. PULSATILLA;

In alternation.

When severe, and complicated with affections of the lungs, indicated by oppressed breathing, pains in the chest, &c., or when the eruption appears to be suppressed, or disappears suddenly, the remedy is

BRYONIA.

When vomiting takes place, attended with great oppression, weakness, and anxiety, there may be given

1 IPECACUANHA; 2. ARSENICUM;

In alternation.

Pulsatilla, however, is the principal remedy in the

13 *

pure form of measles, and will act as a preventive or
modifier of the affection, when administered during, or
previous to, the appearance of the precursory symptoms.

DOSE. — Ten globules, dissolved in four table-spoonfuls of water, and one tea-spoonful given every three
hours.

DIET. — Gruel, barley water, and the like, must be
given until the eruption has declined, when more nourishing food will be needed, as sago or tapioca pudding,
bread, and broth.

SCARLET FEVER. (*Scarlatina.*)

This eruptive fever is, in many districts of our
country, one of the most serious diseases of childhood.
It often appears as a prevalent and malignant epidemic
in autumn, and is, beyond doubt, a very contagious
disease. The eruption usually appears about two days
after the commencement of illness, in minute red points,
first on the face and neck, soon extending over the
whole body in a general deep red blush or efflorescence,
attended with a swollen state of the skin. The fever
manifests itself, like other fevers, by chills, alternating
with heat, lassitude, thirst, dry skin, very rapid pulse,
sometimes with nausea and vomiting; and it continues
generally during the whole course of the eruption,
which disappears at the eighth or ninth day; its disappearance being attended with desquamation, or a scaly
removal of the cuticle. A slight inflammation of the
throat accompanies this fever, when mild; but when it

is severe, this symptom is quite prominent. In such a case, there is present great difficulty of swallowing, with hoarseness, laborious breathing, ulcers in throat, stiffness of the muscles of the neck, and very great constitutional disturbance, with symptoms that belong to the worst description of typhus fever.

Scarlet fever may be distinguished from measles by the more general bright redness and undefined form of the eruption, by the inflammation of the throat, by the deep redness of the tongue, the quickness of the pulse, and by the absence of catarrhal symptoms, as sneezing and cough, that accompany the latter.

The mild form of this disease seldom requires more than a few doses of *Aconite* and *Belladonna*, with proper attention to diet. But its severer form in which the throat is deeply involved is a dangerous distemper, and often proves fatal under the " regular treatment," so called. There is no disease in which homœopathy has more strikingly proved its superiority over all other modes of treatment than in that of scarlet fever.

The medicine that has proved a specific for the pure, uncomplicated form, and which is, in the largest number of cases, almost wholly to be relied on, is

BELLADONNA.

If the fever runs very high, and there is **extreme** restlessness, it would be advisable to make use of **the** following : —

l. ACONITE; 2. BELLADONNA;

In alternation.

Should the throat affection be severe, with ulcers and great accumulation of phlegm in throat, then should be given

1. MERCURIUS; 2. BELLADONNA;

In alternation.

References : —

ARSENICUM; PULSATILLA; SULPHUR.

DOSE. — Ten globules, to be dissolved in a wine-glass full of water, and a spoonful to be given at the end of every hour, every second hour, or third hour, according to the violence of the disease.

DIET. — During the prevalence of febrile symptoms nothing should be given but pure water, thin barley water, or toast water ; and after the abatement of the fever, the return to more solid nutriment must be gradual, the usual diet not being resumed for at least a fortnight, in severe cases, after the disappearance of the eruption.

SCARLET RASH. (*Purpura.*)

This is a distinct disease from the above, though it has, by some, been confounded with it. The eruption is of a dark red or purple hue, and the skin does not turn white on pressure with the finger, as is the case in scarlatina. The surface in the latter is smooth and glossy, while in the present disease, the hand, being passed over the eruption, will detect thickly-studded, small elevations beneath the skin. Fever does not al ways exist with the eruption, and when present, is quite irregular in its course. Soreness in the throat is an accompaniment, and also a moist state of the skin. The specific for scarlatina, viz., *Belladonna*, is not applicable to this disease ; but it is controlled, almost invariably, by, —

l. ACONITE; 2. COFFEA;

In alternation.

Dose. — Eight globules of each, in two wine-glasses full of water, a spoonful to be given alternately every three hours.

Diet. — As in scarlet fever.

MILIARY FEVER.

The characteristic of this disease is an eruption of small, red pimples, which, in two or three days, become vesicular, or like minute blisters, that soon desquamate, or dry in the form of scales, to be succeeded by a fresh crop of vesicles. Like other eruptive disorders, it is preceded and attended by fever, with chills, heat, and marked perspiration, anxiety and restlessness, and a pricking or burning sensation in skin just previous to the appearance of the pimples. It occurs sometimes as a distinct affection, but is more frequently an accompaniment of other diseases. The remedies mentioned in the foregoing article are also the most applicable to this form of fever, viz. : —

l. ACONITE; 2. COFFEA;

In alternation.

When the head is affected, and there is delirium, the *Coffea* should give place to *Belladonna*.

1. ACONITE; 2. BELLADONNA;

In alternation.

DOSE. — As in the preceding disease.

NETTLE RASH. (*Urticaria.*)

This disease is known by an eruption on the skin, like that produced by the sting of the nettle, consisting of elevations, sometimes quite prominent, either white or slightly red, appearing suddenly and changing from one place to another. It arises generally from a constitutional cause, though at times directly dependent on disturbances of digestion from irritating, indigestible food, as spices, pork, sausages, &c. The eating of shell fish has been known to produce it. A precursory fever is present, as in all the previously described eruptions; but it is usually mild, the prominent cause of the complaint being the derangement of the stomach. The first appearance of the eruption is accompanied by heat, itching, and occasionally swelling. The principal remedy is

RHUS TOXICODENDRON.

In case of the sudden disappearance of the eruption from any cause, the medicine to be resorted to here, as well as for the like circumstance in nearly all disorders of this class, is

BRYONIA.

DOSE. — Ten globules dissolved in four table-spoon

fuls of water, one spoonful to be given three times a day, or every six hours; except in case of retrocession, when the dose should be repeated every half hour until the reappearance of the eruption.

DIET.— As in all the eruptive diseases, the presence of fever precludes the use of all solid aliment.

ERYSIPELAS. (*St. Anthony's Fire.*)

This cannot with strict propriety be regarded as an eruptive disease, since its external manifestation is altogether inflammatory; yet the preceding and accompanying constitutional disturbance, together with the general symptoms presented, bear sufficient resemblance to the foregoing to justify its introduction into the present class of disorders.

The fever is sometimes violent, nausea and vomiting, with occasional delirium, attending. The tongue is not dry, as in common fevers, but moist from the commencement, and uniformly covered with a white coating. A florid red hue of the skin is perceived on the third day of illness, with a sensation of heat, and the redness slowly extends, with a slightly swollen border, until the disease is subdued, or the character of the efflorescence has changed. Small blisters or vesicles appear on the inflamed skin about the fourth day, which, in severe cases, sometimes degenerate into obstinate ulcers. When this disease attacks the face, which is its most frequent location, the features are greatly deformed, the eyes closed by the swelling of their lids, and the whole head is often covered by the gradually extending inflammation. As in all fevers of this class, the general disturbance and the local manifestations exhibit every degree of severity; while in all forms of the true erysipelas, from the mildest to the most serious, where the inflammation is confined to the face and head, the chief remedies are,—

1. ACONITE; 2. BELLADONNA;

In alternation.

If the inflammation extends, notwithstanding the above alternation, or if it attacks the joints, and especially if the blistering or vesication is extensive, the medicine to be given is

RHUS TOXICODENDRON.

When wandering from one place to another,—

PULSATILLA.

DOSE.—Ten globules to be dissolved in four table-spoonfuls of water, and one tea-spoonful given every second or third hour.

DIET.—Same as in fevers. Much care should be exercised during the progress of recovery, in guarding against exposure to cold, as relapses are apt to occur from this cause. Dry applications only must be made to the skin; of these the most suitable is the powder of wheat starch or rye meal.

There is a species of erysipelatous inflammation which usually appears on the trunk of the body, and is commonly called " shingles," a term corrupted from the French word " ceingle;" in Latin, " cingulum," signifying a belt. It is composed of clusters of small vesicles, extending sometimes entirely, but seldom more than half round the waist, or across the shoulders. These vesicles, or small blisters, particularly during the process of formation, are accompanied by a sensation of heat and itching, and occasionally by slight general disturbance, as nausea, headache, and want of appetite. The inflammation is induced by sudden transitions from cold to heat, by indigestible food, or intemperance in eating and drinking; but it is attended with no danger, disappearing in ten or twelve days by desquamation of the cuticle. Sometimes, however, a deep-seated pain in the chest or side is complained of, continuing long after the decline of the efflorescence.

Rhus Toxicodendron and *Arsenicum*, three globules of each, (*Arsenicum* in the morning, and *Rhus* at night,) are the remedies suited to this affection. The subsequent pain above mentioned is frequently removed by **Arnica**, three globules every day.

14

CHAPTER V.

CUTANEOUS DISEASES.

IN the last division of the preceding chapter, the cutaneous or skin diseases that are accompanied with fever, and are of short duration, were alone treated of ; and it is now proposed to describe such affections of the skin as are unattended by any decided disturbance of the circulation. Preliminary to such a description, and in order that a clearer comprehension of the character and condition of the said affections may be obtained, it will be necessary to allude briefly to the construction and function of the locality affected.

The integument that envelops the entire human frame is composed of three distinct membranes or layers—the external being the "cuticle," or scarf-skin, the middle one, called the "rete mucosum," and the inner layer, which is the "cutis," or true skin. The first, or the cuticle, is the thin membrane raised from the surface by a secretion of fluid beneath it, under the name of "blister." It contains no nerves or blood vessels, consequently has no sensation, but is merely a transparent incrustation, like varnish, covering the sensitive membranes beneath.

The "rete mucosum" is a net-like membrane, in which is deposited the coloring matter that constitutes the complexion of different races of men; and the "cutis," or true skin, is a strong, fibrous structure, through which the extremities of the blood vessels and nerves pass to expand in intricate communication between it and the "rete mucosum." This distribution of blood vessels, with their attendant delicate nerves, constitutes the sense of feeling, and on the removal of the outer skin, that acute pain or smarting which results from contact with any external irritant, even the air. When such a contact takes place, blood rapidly rushes to this cutaneous network of vessels, called "capillaries," and the little eminences, or "papillæ," which are formed by the termination of the nervous fibres surrounded by these "capillaries," become enlarged and exceedingly sensitive. This enlargement, at times, is so great as to cause, by pressure, an obstruction of the blood in the vessels passing through the cutis, and this gives rise to the mortification that attends a carbuncle, for example. Transpiration, through the skin, of secreted fluid from the blood is constantly going on, sensibly or insensibly; and when the grosser portion of the fluid accumulates, from any retaining cause, it becomes acrimonious, and produces a great variety of superficial eruptions.

Many diseases of the skin are directly dependent on a vitiated or impoverished state of the blood from the use of oily or salted meats, highly stimulating solids or fluids, the abuse of acids, &c.; mental emotions, hereditary tendencies, and other causes, constituting them strictly constitutional affections. Hence the absurdity of attempting their cure by local applications. The possibility even of a transfer of disease from a comparatively safe locality to an internal and vital organ, ought to be sufficient for the instant rejection of the entire class of repellent lotions and ointments.

Physiologically considered, the skin is an organ of sensation, absorption, and secretion, the true offices of which are to afford means of avoiding injurious influences, to appropriate the necessary properties from the surrounding medium, and to remove from the system whatever tends to disturb health and destroy life.

LICHEN. (*Ringworm.*)

This eruption, of which there are many varieties, occurs in the form of dry pimples, and attacks every part of the body. It is attended with a disagreeable sensation of tingling or itching, aggravated at night, or on exposure to heat. The inflamed papillæ, or pimples, continue from eight to twelve days, when they disappear, leaving the surface covered with scurfy exfoliations. The eruption is usually preceded by slight febrile symptoms; and it is not unfrequently the sequel of acute diseases, especially some forms of fever and inflammatory catarrhs.

A mild variety of the lichen is the "*prickly heat*," so called, which is often induced by a sudden exposure to change from cold to heat, and is quite prevalent among new residents in tropical climates.

Another variety is that known as the "*red*" or "*white gum*" — an eruption to which children are subject during the period of dentition.

This eruption also appears on the palms of the hands, or on the limbs, in a confluent form, terminating in an extensive exfoliation of the cuticle, and is known by the common name of "salt rheum."

Nearly all the forms of the eruption may be cured by, —

1. SULPHUR; 2. RHUS TOXICODENDRON; 3. SEPIA;

Administered in the order enumerated, if, after three globules of the first, no improvement is perceptible.

DOSE. — Three globules, dry, morning and evening.

DIET. — As this eruption is often caused and prolonged by improper diet, it is, of course, important to the cure that no irritating or indigestible article of food be used. Neither should any external application be made.

14 *

HERPES. (*Tetter.*)

This order of cutaneous disease consists of vesicles, instead of dry pimples, or pustules; that is, minute blister-like eminences containing a transparent fluid. The eruption appears in irregular clusters of small vesicles, sometimes confluent, sometimes distinct, that fade and fall off in scurfy exfoliations, or form a crust, which, on becoming detached, leaves on the skin a bright red mark. It is accompanied by a burning or tingling sensation, and often by pain. There are several varieties of herpetic eruptions, one of which has been referred to, by the name of "shingles," in the article on erysipelas. It is, however, a true form of herpes, though partaking of the erysipelatous character. Another variety appears on the lips, ("herpes labialis,") in consequence of a cold, or after fevers, the incrustations from which usually become darker than when in other positions. There are several eruptions of this species, named from the locality they occupy; but all partake of the same nature, and are influenced by the same medicines, most of them being greatly modified, if not cured, by one of the following, viz.: —

1. SULPHUR; 2. SEPIA; 3. MERCURIUS.

Dose. — Three globules taken dry, morning and evening, until an appearance of amendment, when the medicine should be discontinued; or, if amendment does not occur within a week, another medicine should be used.

DIET. — The diet, as in all other cutaneous affections, must be confined to simple, unirritating articles of food.

PORRIGO. (*Ringworm.*)

This is an obstinate and contagious eruption, principally occurs in childhood, and consists of circular clusters of yellow points or pustules, attended with inflammation and itching; usually appears upon the head, when it is commonly known as "scalled head." These pustules break, and a yellow or brown crust is formed, which falls off with the hair, and new pustules appear, passing through the same course of increase, maturation, and decline. The consequence of these repeated attacks is a chronic inflammation of the skin, that continues to produce, over the whole head, crops of pustules, constituting an extremely tedious and troublesome eruption. Happily the disease has grown less frequent, as the means of comfort and cleanliness have increased.

The principal remedies are, —

1. RHUS TOXICODENDRON; 2. SILICEA.

DOSE. — The above may be given in dry globules, three at a dose, morning and evening; the first continued every day for a week, and then followed by the second for the same length of time.

DIET. — Dependence must not be placed entirely on medicine. All the hair must be carefully removed by scissors. Bread poultices may be applied and retained during the night, for the purpose of promoting suppuration about the roots of the hair, thus loosening their connection with the skin. All stimulants must be avoided; the diet should be simple and unirritating; and cold food and drink prohibited. Another description of ringworm — classed as a species of " herpes," since the eruption consists of vesicles, instead of pustules — attacks the face, neck, and upper parts of the body, in the form of small rings of vesicles, the portion of skin enclosed within the rings preserving the natural color. The vesicles that form the circumference of the circles break like the preceding, and form brown crusts, but do not reappear in the same place. In most cases, the only medicine needed is

SEPIA.

Dose. — Three globules repeated every other day, until improvement is effected. No outward application need be made.

ACNE

Is a pustular eruption on the face, at or soon after the period of puberty, first appearing as inflammation, on which arise small, distinct pustules.

The same medicines may be used that are mentioned in the preceding section, in the same doses, viz.: —

RHUS, SILICEA.

Diet. — No stimulants of any kind, neither oils, spices, nor salt food, must be taken. There should be frequent bathing of the face in warm water, to which may be added a small quantity of castile soap.

CRUSTA LACTEA. (*Milk Crust.*)

This is an eruption of a similar character with the preceding, but occurs during infancy, first attacking the face, and sometimes extending over the body. It appears in the form of small vesicles, with a red base, which burst and form thin yellow crusts, that are attended with considerable irritation, inducing great restlessness in the child, and causing it to rub frequently -the affected skin. When there is much inflammation, with swelling, and great excitability, relief may be afforded by

ACONITE.

Three globules, morning and evening. After the inflammatory symptoms are lessened, the remedy to be relied on for a cure is

RHUS TOXICODENDRON.

Three globules, morning and evening, until the disease is subdued. No application, except mild soap and water, twice a day, should be made.

PSORA. (*Itch.*)

An eruption of small vesicles, appearing principally on the wrists, between the fingers, and in the flexures of the joints, though existing on any portion of the surface, and, as its common name imports, attended by itching.   It is unaccompanied by fever, arises at times spontaneously among those who are uncleanly in their habits, but is receivable by all persons exposed to the contagion, under any circumstance in life.   A peculiar insect has been discovered in the vesicles; but whether it is the cause or consequence, or whether a necessary connection exists between the insect and the eruption, is a point yet undetermined.   The sudden repulsion of this cutaneous affection has been known to result in various internal diseases, and is believed even to have produced mental derangement.   The founder of homœopathy attributed to improperly-treated psora the existence of numerous chronic disorders.   There is no doubt that this affection of the skin has been occasionally cured by external ointments; but as serious evil has unquestionably arisen from its removal by this mode of treatment, it is obviously the most prudent course to trust to internal remedies, though a longer time may be required for its cure.   The true specific for the uncomplicated form of this eruption is

SULPHUR.

Dose. — Five or six globules may be given, night and morning, for a week or more. If an improvement, however, is not effected in this way within nine or ten days, the tincture of sulphur should be obtained, six drops of which may be dissolved in a wine-glass full of water, and a spoonful of the solution taken night and morning, applying, at the same time, the tincture to the affected skin.

Diet. — As in other cutaneous diseases.

PSORIASIS.

A cutaneous affection, of the same general character with the preceding, under which head it is sometimes included, but differing in being attended with fissures in the skin, and terminating in scales, often confined to an irregularly circumscribed spot — red, itching, and chapped. Its usual location is on the temple, the knee or elbow joint, or the hand. It occurs in various degrees of severity, and in all its forms is generally relieved, if not cured, by

LYCOPODIUM.

Given in the same manner as the *Sulphur* in the preceding article.

BILE. (*Furunculus*)

Is a hard, circumscribed, large-sized pustule, which slowly inflames and suppurates, being attended with much pain. It is common to any part of the surface of the body, and is sometimes productive of considerable febrile irritation. After suppuration, a central, hard substance is thrown out, and the bile disappears, to be followed, perhaps, by others, in different locations.

The medicine that has proved best adapted to promote the healing, and prevent a return, is

ARNICA.

DOSE. — Three globules, morning and evening, with

a weak solution of the arnica tincture applied externally.

CARBUNCLE. (*Anthrax.*)

This is a hard, circumscribed swelling, similar in appearance to the "bile," but larger, more highly inflamed,
and more protracted in its course. The inflammation
terminates in mortification, and small holes are formed
in the centre of the carbuncle, from which issues very
unhealthy matter. This eruption involves the cutis
and the tissue beneath it (the cellular tissue) in inflammation, is extremely painful, and if occurring, as it
usually does, in debilitated constitutions, causes much
disturbance of the system.   The medicine that has been
found most serviceable — sufficient, in fact, to cure,
under favorable circumstances — is

SILICEA.

If this fail to check its progress, and the carbuncle
assumes a dark appearance, enlarging rapidly, it will
be necessary to give

LACHESIS.

DOSE. — Three globules every morning and evening.

A poultice, made of bread and milk, may be applied
to the part affected at the commencement, to relieve
pain, and to promote suppuration.

WHITLOW. (*Paronychia.*)

An extremely painful inflammation of the end of the finger, with considerable swelling, often ending in suppuration, is designated by the above term. The severe pain results from the pressure of the unyielding integuments upon the swelling and inflamed vessels beneath. In the worst form, the inflammation and suppuration .takes place next the bone, beneath the firm tendons. In other cases, the inflammation occurs between the cutis and the cuticle, the pain being less severe than when under the tendons. A bread and milk poultice should be applied, to relieve the tension; and if the inflammation progresses, and is deeply seated, much suffering may be prevented by a free incision, an inch or more in length, made with a lancet, down to the bone. This course is not always necessary, as suppuration may be promoted, and the pain consequently removed, by the application of the poultice, and the administration of the following remedies, viz: —

1. ACONITE; 2. MERCURIUS;

In alternation.

DOSE. — Eight globules, taken dry, every six hours.

CHILBLAINS.

An inflammatory swelling, of a purple color, and quite painful, attacks the extremities, more especially the heel, after exposure to extreme changes from cold

15

to heat. The pain is intermittent rather than constant, and is accompanied by a great degree of itching, while the inflammation at times terminates in ulceration. Children are more subject to this affection than adults, and those of a scrofulous habit suffer severely.

The medicines most efficacious are, —

1. SULPHUR; 2. CALCAREA;

In alternation.

DOSE. — Five or six globules may be taken, in a dry state, every day, first of *Sulphur*, then of *Calcarea* on the succeeding evening, followed by *Sulphur* the next day, and so on, until improvement is perceptible.

CORNS.

These consist of hardened portions of cuticle, produced by the pressure on the foot of a closely-fitting boot or shoe; although similar callosities sometimes appear on prominent portions, as the elbow, knee, or hip, when subjected to long-continued pressure. The parts around the corns are liable to inflammation, by which great suffering is occasioned.

The most successful mode of treatment is to bathe the affected part in warm water for ten or fifteen minutes, then to cut the corn as closely as convenient with a sharp knife, and to rub upon it the *Tincture of Arnica*, diluted with four times the quantity of water. The tendency to inflammation will be thus obviated; and to prevent subsequent growth, a circular piece of buckskin, spread with adhesive plaster, may be applied, having a hole cut in the centre, somewhat larger than the base of the corn. By this arrangement, no pressure can be exerted on the corn itself, and, the exciting cause being removed, the skin will be restored to its natural condition.

WARTS.

These are excrescences formed in the cutis, and are not, as are corns, productions of the outer skin or cuticle. They are often dependent on an unhealthy constitutional condition ; and in order that this remote cause be favorably acted upon, the best medicines to be administered are

1. CALCAREA CARBONICA; 2. SULPHUR;

In alternation.

DOSE. — Six globules of *Calcarea*, to be taken at night, dry, and two days after, a similar dose of *Sulphur*, to be followed in two days by *Calcarea;* and this alternation is to be continued until the wart begins to diminish in size. It is unsafe to apply caustics, as ulceration, and even cancer, has been known to result from their use. The wart may be steamed with the vapor of hot water with safety, and often with benefit.

ABSCESSES.

An abscess is a collection of purulent matter, the result of inflammation either in the substance of some internal organ, or on the surface of the body. When situated externally, — the only form adapted to domestic treatment, — there is, at first, a throbbing pain, with heat and redness; and as inflammation gives place to the formation of matter, the color changes from red to white, a conical prominence appears in the centre of the swelling; the skin over this point, becoming thin, soon gives way, and permits the contents of the cavity to escape. Should not this take place readily, an incision may be made with a lancet. The incision should not, however, be attempted, if a strong pulsation is evident in the tumor, and the conical white projection is not apparent, since it might prove an arterial distention (*aneurism*). When the abscess is of great extent, and especially if situated on the back or thigh, a surgeon's aid should be sought, as no unprofessional person can properly judge of its cause or character, and serious injury might be produced by attempting the removal of a tumor in this locality.

The chief end of treatment consists in the relief of suffering by the promotion of suppuration; and to effect this most speedily, the medicine to be used is

HEPAR SULPHURIS.

Dose. — Three or four globules repeated every three hours.

ULCERS.

These open, suppurating sores upon the surface result from wounds, from certain internal diseases, and from all causes that produce inflammation. While the process of ulceration continues, and is "unhealthy," as it is termed, the surrounding edges of the excavation are inflamed and painful; and a discharge of a corroding nature, thin, watery, and sometimes tinged with blood, takes place. When of a healthy character, or about to heal, red points of flesh, of a cone-like shape, called "granulations," form within the ulcer, gradually filling up the excavation, and rising to a level with the surrounding skin. When these granulations are of a livid hue, and rise much higher than the surrounding surface, they are unhealthy, and do not readily form a new skin. External applications, so much used ordinarily, are not only of little service in general, but frequently interrupt the restorative process. The least injurious ointment is the simple cerate, or the yellow "basilicum" of the shops.

Ulcers are designated by various terms, the enumeration of which would prove of but small practical benefit. The most common and important, however, for which the chief remedies will be mentioned, are the "indolent," the "irritable," "varicose," and "specific."

For the "indolent ulcer," characterized by hard, inverted edges, pale granulations, thick, dark-yellow pus adhering to the bottom of the ulcer, the remedies are, —

1. SILICEA; 2. SULPHUR;

In alternation. Three globules of each every day.

15 *

For the " irritable ulcer," known by its readiness to bleed, its tenderness to the touch, its thin, watery, and sanious discharge, red, swollen, painful, and burning surrounding surface, the remedy is

ARSENICUM.

Three globules given every day.

For the " specific ulcer," so named from being the result of some specific disease, as scrofula, syphilis, &c., the chief remedy is

MERCURIUS.

Three globules every day.

When the veins of the lower limbs become " varicose," that is, distended with blood, they sometimes are ruptured, and the " varicose ulcer " is formed. Bandaging the limb, and giving *Lachesis* and *Pulsatilla*, — three globules of each in alternation every second day, — is the course recommended.

It is sometimes advisable, while the suitable medicine is being taken internally, that a solution of the same (ten or twelve globules in a wine-glass of water) be made for external application. Soft cloths dipped in cold water, and kept constantly wet, are the best dressings for ulcers in general; while once a day they may be wet with the medicinal solution above referred to.

CHAPTER VI.

DISEASES OF THE EYE AND EAR.

THE complex and very delicate organs of sight and hearing are subject to a variety of affections, many of which are extremely painful; some threatening the destruction of those organs, and even of life itself. The severer forms of disease, involving the internal structure, as deep-seated inflammations, morbid growths, &c., should be submitted to the attention of a professed aurist or oculist; but affections of the outer membranes, injuries from external irritants, superficial inflammations, may properly form the subject of unprofessional study. The most common affection of the eye is an inflammation of its outer coverings, or, as it is termed,

OPHTHALMIA.   (*Inflammation of the Eye.*)

The inflammation commences with a sensation as of some foreign substance between the eye and its lid, soon followed by heat, redness, swelling, and intoler-ance of light, with some degree of pain in the eye and nead, together with more or less sympathetic febrile action.   When the inflammation is deep seated, there are severe shooting pains through the head, and con-siderable constitutional disturbance, with a plentiful flow of tears ; and during the progress of the inflamma-tion, a viscid secretion adheres to the lids, adding much to the inconvenience and suffering.

Ophthalmia is induced by cold, continued exposure to a very strong light, local injury, and, in certain seasons and climates, from some peculiar condition of the atmosphere, it prevails as an epidemic.

For all degrees of this affection of the eyes, the chief remedies are, —

1. ACONITE ; 2. BELLADONNA;

In alternation.

DOSE. — Four globules of each, every three hours, or more or less frequently, in proportion to the degree of inflammation.

When the result of external injury, water, in which ten or twelve globules of *Aconite* are dissolv d, should be applied as a wash, together with its internal ad-ministration, as above directed.   The same means are

to be adopted when ophthalmia arises from the presence of any foreign substance in the eye, after the
removal of the offending cause. Should the eyelid be
injured, or both the eye and lid be bruised, *Arnica*
may be used in the same manner.

As previously stated, there are deep-seated diseases
of the eye, for the cure of which other remedies are to
be used; but it is not necessary to detail the treatment of these diseases, as sufficiently accurate information cannot be conveyed through a popular treatise;
furthermore, the chronic character of such affections
will enable one to await the procurement of the best
advice.

DIET. — The food should be simple and unirritating.
If much fever is present, the different kinds of gruels
must comprise the entire diet. The eyes should be
shaded from light and heat, but by no means tightly
bandaged. Cold water may be freely used as a wash,
both for the head and eyes.

STY. (*Hordeolum.*)

This term is applied to a small, hard, deep-red tumor, situated on the eyelids, that is often quite painful, and is occasionally attended with considerable
inflammation, which spreads over the outer covering
or conjunctiva of the eye. It is caused by an obstruction of the outlet of one of the small glands of the
lid, by reason of which the secretion furnished by
the gland is arrested and forms the sty. Sometimes

much febrile action is excited by its presence in children.

The remedies are, —

1. ACONITE; 2. PULSATILLA;

In alternation.

DOSE. — Six globules of each of the above are to be given, dry, every day; the first in the morning, the second in the evening.

Should an indurated spot remain, after the inflammation has subsided, or should a disposition to a succession of stys be manifest, the remedy is

CALCAREA CARBONICA.

Six globules given every other day.

STRABISMUS. (*Squinting.*)

This affection, by which objects are seen in an oblique manner, the axis of vision being distorted, commonly occurring in childhood, and frequently becoming permanent, may generally be remedied by the following method: Place the child so that the light may fall equally upon both eyes, carefully avoiding that position which will force one eye to more constant use than the other. Should one eye be already turned aside, the opposite one may be covered with a shade for several days, so that the squinting eye will be used exclusively, until direct vision is restored. If both eyes turn outwards, some dark substance placed upon the tip of the nose will often remedy the defect; or if inwards, two pieces of stiff pasteboard may be so bound on each side of the head as to project beyond the level of the eyes in front.

If the defect be produced by other means than obliquity of light, medicinal remedies may be resorted to. If there is much accompanying heat in the head, with throbbing of the arteries, give

BELLADONNA.

Should the foregoing be unsuccessful, and there is reason to suspect the existence of worms, give

HYOSCYAMUS.

DOSE. — Six globules of the medicine selected should be given in a dry state every day.

DEFECTS OF VISION.

When the eye projects more than is natural, or when the crystalline lens within the eye is too convex, the rays of light passing through the lens are converged to a focus before reaching the retina, and distant objects are indistinctly seen. This defect of vision is called near-sightedness, and concave glasses are made use of to counteract such excess of convexity. When, on the contrary, the lens or cornea is not sufficiently convex to draw the focus of rays to the proper point, that is, to the retina, a near object cannot be seen distinctly. In this case, the focus falls beyond the retina, and the assistance of convex glasses is consequently needed. This loss of rotundity is usually owing to the absorption of substance as age advances, and is termed far-sightedness. The medium, however, through which the rays of light pass may be affected by disease, either local or sympathetic.

NEAR-SIGHTEDNESS.

If occurring in consequence of inflammation of the eyes, and not of long duration, the medicine that has been given with good effect, is

PULSATILLA.

If brought on by the use of mercury,

CARBO VEGETABILIS.

DOSE. — Six globules, daily.

FAR-SIGHTEDNESS.

When, from other cause than the gradually lessening convexity of the eye by absorption, a difficulty in perceiving near objects is experienced, a medicine that has been found beneficial is

DROSERA.

If no good effect is perceived from the above, it may be followed by

SULPHUR.

DOSE. — Same as above.

WEAKNESS OF SIGHT.

When this condition exists, the following medicines will prove beneficial, under the circumstances indicated, viz. : —

PULSATILLA.

When attended with general debility, and frequent lachrymation, the pupil being of a light gray hue.

SULPHURIC ACID.

Burning pressure in the eye, with cloudy vision; irritability; emaciation; or when consequent on the sudden disappearance of hemorrhoids.

RHUS TOXICODENDRON.

When occurring soon after the sudden disappearance of any cutaneous eruption, of rheumatism, or of gout.

DOSE. — Six globules, daily.

No ointments or astringent washes should be applied to the eyes in any case. Pure water, either cold or warm, as may be most agreeable, is the best application that can be made.

AMAUROSIS.

In this affection of the eye, there is a diminished or total loss of sight, without any apparent alteration externally. The optic nerve, which proceeds from the brain to the eye, expanding into the membrane called the retina, is in a paralytic condition ; and this state is brought on by causes that debilitate the system, or by violent exertion, that determines blood to the brain, by mental emotions, by the use of narcotic and other drugs, or by continued exposure to a strong light. It may also be caused by morbid growths within the cranium compressing the optic nerve. Slight amaurosis sometimes attends stomachic and menstrual derangement, nervous and intestinal irritation.

The indications of its approach are, usually, appearances as of cobwebs or small dark specks floating
before the eyes; and sometimes flashes of light, of
various colors, changing into a gradually condensing
mist, and terminating in partial or entire obscuration
of sight. At times, the disease progresses rapidly; so
that, in the course of a few days, light cannot be distinguished from darkness. In such cases, it is more
easily influenced by medicine than when the advance
has been gradual; since, in the latter instance, it is
presumed to depend on constitutional derangement.

The chief remedies are, —

1. BELLADONNA.

Dilated, immovable pupil; the flashes of light, or
floating particles, appear red; headache; sense of
fulness of, or pressure on, the eye; heat and pain.

2 HELLEBORUS.

Extreme sensibility to light; aching pain passing
from the back of the head to the forehead, attended
with catarrhal symptoms.

3. NUX VOMICA.

Sensibility to light more decided in the morning,
with frequent contraction of the eyelids; when
amaurosis results from indigestion, or the use of
stimulants.

Reference : —

CHINA; PULSATILLA; PHOSPHORUS.

DOSE. — Six globules dry, every evening, discontinuing as soon as improvement is apparent. If the improvement is not progressive, the same medicine should be resumed.

DIET. — All heating, irritating articles of food must be avoided, and stimulation of every kind refrained from.

FOREIGN BODIES IN THE EYE.

Substances of different sizes and degrees of solidity insinuate themselves under the eyelids, occasioning considerable irritation, and sometimes great pain. If the irritating body be floating about on the surface of the eye, it can be easily seen and removed by a strip of paper rolled up in a conical form, the apex of the cone being moistened and brought in contact with the substance. If the body be minute, and under the upper eyelid, by placing a wooden netting needle or a similar object across the lid, the latter can be drawn back over it and wholly inverted, so that the smallest particle may be perceived and withdrawn by the moistened paper. The under lid is easily depressed by the finger and the lower half of the eyeball exposed to view. In case a metallic particle is thrown into the eye so forcibly as to become firmly embedded in its surface, the assistance of a surgeon should be obtained, as rough or awkward management may deprive the person of sight. If loosely adhering to the surface, a pair of small pincers will be sufficient to detach the particle, without injury to the eye, if carefully used by any one. The subsequent inflammation sometimes caused may be removed by frequently bathing the **eye** in tepid water, and taking every two hours six globules of *Aconite*.

If lime, or any cement of which lime is an ingredient, enters the eye, a mixture of vinegar and water, in the proportion of one part of vinegar to ten parts of water, must be frequently used as an external application.

DISEASES OF THE EAR

Are generally of an intricate nature, and not readily adapted to popular treatment; but a description of the external and most common affections of the ear will not be out of place here, while it is to be understood that such as are obscure and complicated must be submitted to the attention of those who make diseases of this organ their special study.

It not unfrequently happens that obstinate pains in the ear, unattended by any other local symptom, and produced by no directly obvious cause, exist as a sympathetic consequence of defective teeth or diseased gums. It is always proper, therefore, in such cases, to have the mouth thoroughly examined.

INFLAMMATION OF THE EAR. (*Otitis.*)

This is known by feverish symptoms, an excruciating pain within the ear, and sometimes a wild manner and incoherent speech — in short, delirium — which is the effect of inflammation existing so near to the brain. The external ear is red, hot, swollen, and exceedingly sensitive to the touch. The pain is of a burning, throbbing, piercing character. The medicines which will relieve this distressing affection are, —

1. ACONITE; 2. PULSATILLA;

In alternation.

DOSE. — Ten globules of each are to be dissolved in two table-spoonfuls of water, and one tea-spoonful of each given every hour, commencing with the first, one hour afterwards giving the second, the next hour the first again, and so alternating till relieved.

A sponge dipped in warm water, and bound on the ear, has been recommended in preference to a poultice or other applications.

EARACHE. (*Otalgia.*)

This pain is often unattended by any symptoms of inflammation, and may be the result of a cold, an affection of the nerves, of rheumatism, the repercussion of an eruption, and other causes; but, however caused, a remedy may in most cases be found by reference to the following indications : —

1. PULSATILLA.

Jerking, tearing pain, extending to the side of the face; external redness, heat, and swelling; applicable to mild, melancholy temperaments.

16*

2. BELLADONNA.

Pricking, shooting pain, with a rumbling sound; head and eyes affected; pain increased by touch **or** motion.

3. MERCURIUS.

Burning pain externally, with a sensation of cold internally; severe internal pain, with violent twinges; increased pain from warmth; discharge from the ear.

4. CHAMOMILLA.

Stabbing pain, as from a knife; great sensitiveness to noise; excitability; fretfulness.

References: —

ARNICA; NUX VOMICA; SULPHUR.

DOSE. — Ten globules dissolved in four table-spoonfuls of water, and a tea-spoonful given every one, two, or three hours, according to the severity of the pain.

A discharge from the ear, which is usually the result of inflammation, may be treated by

1. PULSATILLA; 2. BELLADONNA;

In alternation.

DOSE. — As above.

DEAFNESS

May arise from various causes, some occasioning temporary, others permanent and incurable deafness. Very loud sounds, severe colds, paralysis of the auditory nerves, inflammation of the membranes, accumulation and hardening of the ear secretion, the incautious use of earpicks, original malformation, and certain fevers may be enumerated as causes of both kinds. These are not always easily traceable, neither is the true condition readily discovered, and there are but few reliable indications; nevertheless, under certain circumstances, remedies are designated and available.

If, for example, deafness follows an attack of scarlet fever, the medicine almost invariably serviceable is

BELLADONNA.

If in consequence of small pox,

MERCURIUS.

If in consequence of measles,

PULSATILLA.

If after the sudden disappearance of an eruption,

SULPHUR.

If after fevers, especially nervous, and the deafness is attended with a sensation of fulness in the head,

PHOSPHORUS.

If from cold, or a transfer of rheumatic pains to the head,

DULCAMARA.

If from a hardened secretion, syringing the ear with warm milk and water will remove the cause.

DOSE. — Of either of the above, six globules, dry, morning and evening.

FOREIGN BODIES IN THE EAR.

When insects obtain an entrance into the ear, a sure mode of dislodging or destroying them is to pour a few drops of sweet oil into the ear, the head being retained in a suitable position for the purpose.

When any solid substance is forced into the ear, great care must be exercised in removing it, as much injury might be caused by misdirected attempts to effect such removal. If the body present a surface that may be firmly grasped by a pair of pincers, its removal will be easily accomplished; but when the substance is round and smooth, like a small marble, for example, any attempt to extract it in the above way would only result in embedding it more deeply. The best method is to insert gently, between the ear and the substance, the doubled end of a very small wire, bent on itself, like a lady's hair pin, that the substance may be acted upon from behind, and thus removed. The injurious results of the continued pressure of a foreign body in the ear, or of careless efforts for its extraction, would be inflammation of the tympanum, or drum of the ear, and perhaps its consequent destruction by ulceration.

CHAPTER VII.

GENERAL DISEASES.

RHEUMATISM.

INFLAMMATORY or Acute Rheumatism is generally manifested, at its commencement, by the usual symptoms of fever, as white-coated tongue, quick, full pulse; thirst, lassitude, chills, followed by heat; and soon after, aching, darting, or throbbing pains are felt in the joints, with heat, redness, and swelling at the seat of the pain. Perspiration often succeeds or accompanies the feverish symptoms, but does not, as is usual in fevers, bring relief, and is no sign of convalescence. The pain, which is principally confined to the joints, though occasionally occurring in the neck, chest, or head, is transferred from one place to another, and is greatly increased by exposure to cold air, and by motion. When a transfer of the disease takes place from the joints to an internal organ, as the heart or brain, recovery is doubtful. This transfer, however, is rare, and, in its usual course, the disorder is not attended with danger, although quite protracted and painful.

Rheumatism frequently occurs in a chronic form, in which case there are no indications of fever, no signs of local inflammation, but fixed, dull pains in the joints, with muscular rigidity, weakness, and numbness.

The cause of rheumatism is, chiefly, obstructed perspiration from exposure to cold when the body is heated, or from dampness of clothing, wet feet, &c. It prevails in our climate mostly during spring and autumn, and attacks persons of every age and condition.

In the treatment of this affection, *Aconite* should always be first given, as usual, for the mitigation of fever, after which the following, as indicated, viz : —

l. BRYONIA.

Shooting pains, increased by motion or cold, aggravated at night, or on rising; pain in the loins or chest, irritability, difficult breathing.

2. RHUS TOXICODENDRON.

Sensation of torpor or numbness; weakness of the limbs; pain aggravated by rest; relieved by motion; pain as from a bruise.

8. PULSATILLA.

Shifting pains, worse in the evening. In chronic rheumatism, when there are muscular rigidity and numbness.

4. BELLADONNA.

Severe burning pains, increased by touch ; great nervous excitement, with tendency of disease to the head, indicated by flushed face, throbbing of arteries of head and neck.

References : —

CHAMOMILLA; NUX VOMICA; MERCURIUS; SULPHUR.

DOSE. — The medicine selected should be dissolved in the proportion of one globule to two tea-spoonfuls of water, and one tea-spoonful given every three to six hours, according to the severity of the symptoms.

A form of rheumatism, which consists of a fixed, violent pain in the loins, affecting the muscles of that region, sometimes extending upwards to the shoulders, is called " lumbago." It is usually attended with much febrile action, and should be treated by the remedies indicated as above for rheumatism in general.

Another variety consists in a sharp pain centring about the hip joint, often extending to the knee and foot, and termed " sciatica," on account of its being confined to the course of the " sciatic nerve." For the treatment of this form, also, the above medicines should be consulted, and administered as there de scribed.

GOUT. (*Podagra*.)

In many respects, this disease resembles the preceding, and they are not always easily distinguished from each other; in truth, there is often a combination of the two, forming "rheumatic gout." The nature of the inflammation of gout and rheumatism is somewhat similar, and the comparative violence and character of the pain have been humorously described in the following way, by one who appears to have suffered from both : " Place your limb in a vice, and turn the screw until pain becomes intolerable. That represents rheumatism. Give another twist to the instrument, and you will find out what gout is."

The real differences, however, between gout and rheumatism are these : In rheumatism, the large joints are affected; in gout, the smaller joints, and especially of the great toe. The former attacks many parts at the same time ; the latter one only. There is more swelling in gout than in rheumatism, and the paroxysms of pain in the former are of shorter duration. Perspiration relieves the gout, but not the rheumatism. A fit of the gout commences in the night, and there occurs a distinct remission, usually in twenty-four hours. Gout is frequently, not invariably, preceded or accompanied by dyspeptic symptoms, and is more liable to seize upon hearty eaters and intemperate persons than the abstemious and laboring class. As in the case of rheumatism, if confined to the joints, gout is not at all dangerous ; but if translated to some internal organ, the disease assumes a serious aspect.

In the treatment of gout, the chief remedies are, —

l. ACONITE; 2. NUX VOMICA;

In alternation.

If relief is not afforded after two doses of each of the above, then the following are to be given : —

1. ARNICA; 2. PULSATILLA;

Also in alternation.

Reference : —

BRYONIA; ARSENICUM.

DOSE. — Twelve globules, dissolved in a wine-glass of water, and a spoonful given every two hours.

If a transfer should take place to the head, give *Belladonna*, at intervals of an hour, dissolved as above.

If to the stomach, *Nux Vomica*, used in the same manner.

SCURVY. (*Scorbutus.*)

This disorder chiefly arises from a deficiency of vegetable acid in food, in connection with habits which depress the nervous energy, as indolence, confinement in damp air, neglect of cleanliness, fatiguing exertions, and despondent moods of mind. It is well known that seamen are particularly subject to this disease, in con-

17

sequence of being deprived of vegetables, and restricted to the use of salted provisions during a long confinement on shipboard. The cure is always greatly facilitated, if not directly effected, by a change of diet; that being made to consist in part or chiefly of native acids, as the juice of lemons and other fruit, or of vegetable substances that have undergone the acetous fermentation.

The symptoms peculiar to this disease are, great prostration of strength, paleness and bloatedness of the face, sponginess and swelling of the gums, and of the lower limbs; ulcers, and dark-blue spots on the skin, with offensive excretions. The scurvy is of a putrid nature, but has not been satisfactorily proved to be contagious.

Although vegetable acids and fresh provisions are to be mainly relied on for the removal of a condition resulting from the absence of such aliment, the morbid state may, at times, continue in its manifestations of ulcerations, diseased gums, &c., until counteracted by medicinal agents; and those which have proved the most beneficial are, —

1. CARBO VEGETABILIS; 2. ARSENICUM;

In alternation.

Should the action of the above not prove completely curative, the remains of the complaint may be removed by, —

1. SULPHUR; 2. SULPHURIC ACID;

In alternation.

DOSE. — Six globules of each may be given **every day,** the first in the morning, the second in the evening.

*

DROPSY. (*Hydrops.*)

The above name is given to an unnatural accumula-tion of fluid in the cellular tissue that lies between the skin and the muscles, or in the different serous cavities of the body. As this is a disease that is usually symp-tomatic of some disordered condition existing under a great variety of circumstances, and in different loca-tions, and as its treatment cannot be judiciously con-ducted except by those who have had much experience in the management of diseases, it would be foreign to the plan of this work to enter into all the details respecting it. Its chief varieties will therefore be merely referred to, and the important remedies added that are applicable to dropsy in general.

When fluid is effused into the cellular membrane that is situated beneath the skin, it usually extends throughout that tissue, occupying the whole trunk and limbs, and is externally manifested by a soft, inelastic tumefaction. This is termed " anasarca." When the fluid is confined to the abdominal cavity, it is called " ascites." When an effusion of fluid takes place in the cavity of the chest, it receives the name of " hydro-thorax ; " when in the head, " hydrocephalus."

The puncturing of cavities by sharp instruments, for the removal of the accumulated fluid, is often resorted to, but it is a mere palliation, the diseased action no↓

being reached in this manner. The fluid should be removed, if possible, by absorption, and the healthy condition of the secreting and absorbing vessels restored, which may be accomplished the most frequently by the administration of the following, viz : —

1. HELLEBORUS.

Great debility ; febrile symptoms ; shooting pains in the extremities.

2. ARSENICUM.

Difficulty of breathing; general coldness, with thirst.

3. MERCURIUS.

Oppression of the chest; cough; much heat, and perspiration.

Reference : —

DULCAMARA; CINCHONA; PHOSPHORUS.

Dose. — Six globules, every morning and evening.

HYDROPHOBIA.

This name, compounded of two Greek words, signifying " dread of water," is well known as applied to the distressing disease that results from the bite of a rabid animal. The same dread of water, however, with convulsions, excessive restlessness and anxiety, thirst, laborious breathing, and other symptoms, which arise from the absorption of poisonous saliva, has been known to occur spontaneously, when no wound from any animal had been received. This circumstance, together with the fact that, in many cases of dissection, no diseased condition of the internal organs could be discovered, has led to the supposition that the nervous system is principally acted upon in hydrophobia. The throat of a person suffering from this disorder is invariably affected; and it is said that the nearer the wound is to this part, the more dangerous it becomes.

For the treatment of wounds made by any rabid animal, Dr. Hering confidently recommends the application of dry heat. " Whatever," he writes, " is at hand at the moment, a red-hot iron, a live coal, or even a lighted cigar, must be placed as near the wound as possible, without, however, burning the skin, or causing too sharp a pain; but care must be taken to have another instrument ready in the fire, so as never to allow the heat to lose its intensity." It is asserted, also, that the heat should not exercise its influence over too large a surface, but only on the wound and parts adjacent. If oil or grease can be readily procured, it may be applied around the wound; and this operation is to be repeated as often as the skin becomes dry; soap or saliva may be applied when oil or grease cannot be obtained. Whatever is discharged, in any way, from the wound, ought to be carefully removed. The application of burning heat should be continued in this manner until febrile chills affect the patient, and this practice is to be repeated three or four times daily, till the wound is healed without leaving a colored scar. At the same time, the patient should take, every five or six days, or as often as the aggravation of the wound requires it, one dose of *Belladonna*, till the

17 *

cure is completed. If, at the end of seven or eight
days, a small vesicle shows itself under the tongue,
with feverish symptoms, it will be necessary to open it
with a lancet, or sharp-pointed . scissors, and to rinse
the mouth with salt and water.

SEASICKNESS.

In case of an anticipated exposure to this distressing
affection, which is too well known to require a description, six globules of *Nux Vomica*, taken seven or eight
hours previous to embarkation, will lessen the chances
of an attack. Should no such preventive means have
been adopted, the alternation of *Nux Vomica* with *Arsenicum*, at intervals of an hour, in the above dose,
when sickness comes on, will relieve, if not remove
it. Should nausea nevertheless continue, the following medicines may be tried, in the order enumerated, viz. : —

1. ARSENICUM; 2. COCCULUS; 3. IPECACUANHA.

Dose. — Six globules, every two hours.

A too hasty repetition of the dose, as well as a frequent change of remedies, will, in the case of seasickness, as in all other affections, diminish the chances of
relief.

Diet. — No one should venture on the water, if
liable to seasickness, without partaking, an hour or
two previously, of a light meal, consisting of plain,
digestible food. After the first attack of sickness is
over, and the appetite is wholly or partially restored,

the resumption of the usual diet should be preceded by a meal of toasted bread, or ship-biscuit, with coffee, or, if at hand, London porter. Unnecessary as this course may appear, the writer's experience enables him to assert that much suffering will certainly be thereby prevented.

For sickness, arising from the motion of a carriage, the above medicines are equally applicable.

SCROFULA. (*King's Evil.*)

This term is derived from " Scrofus," the Latin for swine. It is a disease to which that unclean and disgusting animal has always been subject, and which, no doubt, led to the prohibition of its flesh, as an article of food, among the ancient Jews. Scrofula is inherited and ingenerated by man, remaining latent until developed by favoring circumstances, as various unhealthy habits of life, cold, damp air, close confinement in ill-ventilated rooms, improper food, and inattention to cleanliness. Its usual earlier external manifestations are a hard, tumefied state of the glands about the neck and upper part of the body, with a ready swelling of the large joints. It is often brought into activity by severe eruptive diseases, improperly treated, and by injuries, as wounds, fractures, &c. ; and it may attack the eyes, the hip-joint, the lungs, and mesenteric glands; appearing, in short, in a great variety of shapes and locations.

The medicines most beneficial in checking the devel

opment, or destroying the scrofulous " diathesis," **as it
is termed,** as well as in healing the suppurating glands,
are, —

1. SULPHUR; 2. CALCAREA CARBONICA;

In alternation, every day ; the first on one day, the
second on the following.

Other medicines have been used with success in cer
tain instances, as *Arsenicum, Mercurius, Silicea, Calcarea,* and *Lycopodium ;* but as the individual cases and
circumstances are so numerous and peculiar, it would
be impossible to designate, within our assigned limits,
every special indication, and as the disease is of a
chronic, constitutional nature, and its true character
and extent difficult to determine in all its variety of
manifestations, professional advice may, and should be
obtained.

DIET. — The food should be nutritious, consisting of
digestible meats, and farinaceous substances, as bread,
potatoes, rice, &c., together with an abundance of un-
adulterated milk.   Pork, and its different preparations,
should be rigidly excluded.   Fresh air and exercise
are of the utmost importance.

RICKETS.   (*Rachitis.*)

This is a disease of the bones, occurring in scrofulous
children, and commencing during infancy by a gradual
distortion of the bones, enlargement of the joints, flac-
cidity of the muscles, incurvation of the spine, enlargement of the head and abdomen, debility, and general

emaciation. Rachitis prevails especially among those children that are poorly nourished, poorly clad, and neglected.

The remedies and regimen advised for the above described disease of scrofula are in every respect applicable to this.

BLEEDING FROM THE NOSE. (*Epistaxis.*)

This hemorrhage is not often of sufficient consequence to excite alarm, and is, at times, rather beneficial than otherwise, particularly in plethoric persons, and at the period of puberty. But when the flow of blood is profuse, and its return frequent, or when it happens to aged or debilitated persons, measures should be taken to arrest it. These consist of external applications, such as sprinkling cold water in the face; applying a piece of ice or a cold metallic substance — as a key, for example — to the back of the neck; wearing a cloth, wet with cold water, round the body; bathing the head in iced water, or the feet in warm water. And of internal remedies, when there is much fever, with flushed face, pulsating arteries of the neck, indicating over-fulness of the vessels of the head, the following, viz. : —

1. ACONITE; 2. BELLADONNA;

In alternation, every half hour.

Reference : —

ARNICA; RHUS TOXICODENDRON; BRYONIA.

The remedy for the state of debility that results from hemorrhage of any kind is

CHINA.

DOSE. — Six globules, every half hour, or ten globules, dissolved in two table-spoonfuls of water, one tea-spoonful of which is to be given every half hour. One dose (six globules) of the latter, viz., *China*, may be sufficient; if not, it should be repeated in three hours.

INFLAMMATION OF THE KIDNEYS. (*Nephritis.*)

The symptoms of this inflammation are acute pains shooting from the lumbar region or loins downwards, with numbness and spasms of the thigh and foot, colic pains, frequent and difficult urination, fever, nausea, and vomiting. This may be distinguished from lumbago, or rheumatism in the lumbar region, by the pain not being much increased by motion, as is the case in the rheumatic affection, by the presence of fever, by nausea and vomiting. Nephritis is induced by external injuries, by violent exercise in riding and walking, by lifting heavy weights, by exposure to cold, the use of stimulating diuretics, and poisoning from Spanish flies.

The chief remedies for this affection, and which will, in most cases, afford relief, are

ACONITE; CANTHARIDES

In alternation.

When the exciting cause has been an injury from a fall, blow, or strain, the remedy is

ARNICA.

Reference: —

NUX VOMICA; BELLADONNA; CANNABIS.

DOSE. — Ten globules dissolved in two table-spoonfuls of water, one tea-spoonful being given every hour until improvement commences, when the interval should be lengthened to two, three, or four hours.

DIET. — By one subject to attacks of the above, no stimulants of any kind should be used. During the violence of the attack, frequent applications of cold water over the seat of the pain will afford much relief.

INFLAMMATION OF THE BLADDER. (*Cystitis.*)

Inflammation of the bladder is indicated by symptoms similar to those mentioned in the preceding article, excepting in the location of the pain, which is in front, over the region of the bladder, and is of a burning character. The remedies above mentioned are applicable to the present form of inflammation, viz.: —

1. ACONITE; 2. CANTHARIDES;

In alternation.

Dose, diet, and regimen, as above directed.

There are other disordered conditions in the class of urinary complaints, which are so complicated and obscure in cause and character, that a particular description will not be attempted here, as the physician's presence is usually indispensable. One, however, that is the most painful, and to which it may be important to refer, as often yielding, when not the result of mechanical obstruction, to our remedies, is a partial or total urinary suppression. In such cases the remedies for " nephritis " may first be tried, viz. : —

1. ACONITE; 2. CANTHARIS;

Alternately, followed by *Nux Vomica* and *Pulsatilla*, if needed.

DIET. — .Warm, mucilaginous drinks, as an infusion of flaxseed, gum arabic, pearl barley, may be frequently taken, and warmth applied externally. All stimulants and animal food should be avoided.

HIP JOINT DISEASE.

This generally occurs in children, though occasionally seen in adults and even aged persons. Its first indication is lameness, with pain in the knee of the affected side, and inability to bear the weight of the body. The disease consists in an inflammation of the hip joint, and ulceration of its cartilages, which, progressing, dislocates the head of the thigh bone from its cavity, producing deformity and shortening of the limb. The symptoms above mentioned are first noticed, and as the disease advances, a fixed pain is felt in the hip, increased by pressure, and a prominence may be noticed at the hip joint where a depression naturally exists. The complaint is usually of a scrofulous nature, and is dangerous in its advanced stage on account of the extreme constitutional disturbance in- duced. It may be arrested at its commencement, but the whole course of treatment demands the skill of a properly educated person. The medicines on which the principal dependence is to be placed are,—

1. MERCURIUS; 2. SULPHUR; 3. CALC. CARBONICA.

DOSE. — Six globules, daily.

WHITE SWELLING.

This term is applied to a chronic inflammation of the knee, with swelling, and the effusion of fluid in and around the joint; it being, like the above, a dis- ease chiefly connected with a scrofulous habit. When appearing suddenly, in consequence of a blow or fall, and not dependent on a scrofulous condition, the fluid is absorbed on the subsidence of inflammation, and the joint is restored to its natural state. But in other cases a permanent swelling and stiffness remain, and a cure is not easily effected.

The chief remedies are,—

1. PULSATILLA; 2. SILICEA; 3. SULPHUR.

DOSE. — As in the foregoing.

In both of the above named diseases, whether of scrofulous origin or otherwise, it is important that absolute rest in a horizontal position be strictly main- tained.

18

HERNIA. (*Rupture.*)

This intestinal protrusion makes its appearance as a tumor in different parts of the abdomen. Without entering into an anatomical description of the parts involved in a rupture, we shall refer at this time only to its causes, characteristics, and treatment. It may be occasioned by blows, falls, violent muscular exertions, and it is sometimes congenital. It is to be distinguished, wherever it appears, from other enlargements by its varying size, being larger in an erect posture, and smaller in a recumbent position, and by its yielding to compression. It is often easily pushed back or reduced, and at times irreducible. When reducible, it is to be returned to its natural place, and retained there by an instrument called a "truss," so contrived as to exert constant and equal pressure on the part.

In the reduction of hernia, the patient is to be placed in a horizontal position, with the head and chest raised, and the lower limbs flexed upon the abdomen, in order that the muscles may be as much relaxed as possible. The protrusion is to be gently and gradually pressed upon, at a moderate elevation, until it has altogether disappeared. When it will not yield to pressure in this way, the application of very cold water will facilitate its return. In case of the impossibility of reduction by the hand, and when it has existed long and become "strangulated," as it is termed, surgical skill is required. The operation necessary, in extreme cases, need not be here described, as it cannot prove of any practical advantage.

In case of a rupture in any situation being painful or very sensitive, and previous to attempts at reduction by the hand, medicines should be given. If incapable of themselves to restore the protruded intestine, they will add to the ease and facility of the operation by diminishing the local irritation, counteracting general febrile disturbance or difficulties of digestion.

The principal remedies are, —

1. ACONITE.

For feverish symptoms, with inflammation at the place of rupture; vomiting and restlessness.

2. NUX VOMICA.

Oppressed, laborious breathing; when caused by indigestion, exposure to cold, or violent mental emotion.

Should neither of the above prove efficacious, they should be followed by

OPIUM.

Dose. — Six globules given every hour.

CHAPTER VIII.

DISEASES OF INFANCY.

1. Colic.
2. Sleeplessness.
3. Milk Crust.
4. Convulsions.
5. Difficulties of Dentition.
6. Vomiting.
7. Inflamed Eyes.

8. Tooth Rash.
9. Worms.
10. Diarrhœa.
11. Cholera Infantum.
12. Aphthæ.
13. Catarrh.

In early life diseased conditions exist that differ in many respects from those incidental to a more fully developed stage of existence, and particular attention to such morbid conditions, and the means whereby they may be distinguished, is, in consequence, required. More especially is this necessary on account of the absence of a well-defined character of disease in the rapidly changing and extremely susceptible infant organization, and of the incapacity, in the very young, of expressing their sensations, by which we should be enabled to judge of the nature and seat of pain, &c. The diagnosis or discrimination of one disease from another, in children, is to be obtained only by observation of external manifestations; and it will be necessary to preface the description of such diseases by a brief allusion to those manifestations that are, in their different forms and under various circumstances, presented. It may be proper to add here, that by reason of the irritable condition and great susceptibility of young life to influences of every kind, the gentle action of "infinitesimal doses" is, in an eminent degree, adapted to it remedially, and the serious results of the energetic and uncertain medication of the "old school" in disorders of infancy never appear more sadly prominent than when contrasted with the happy consequences of homœopathic management.

The physical symptoms by which diseases of young children are recognized consist of the position of the limbs or trunk, the temperature of the surface, the state of the secretions, the pulse, the tongue, the respiration, in addition to one that is of the first importance, as it is a very general manifestation of suffering, namely, *crying*. Whenever this is attended by a restless, uneasy condition of body, it is a sign of pain or some uncomfortable sensation; when accom- panied with drawing up of the lower limbs on the abdomen, a colic pain is indicated; when the hands are frequently thrust into the mouth, and bitten, there exists pain from dentition; when crying is attended and preceded by cough, pain in the chest is present.

The state of the respiration is a valuable means of diagnosis. Frequent and rapid breathing, especially

18 *

if attended by a dry cough, indicates inflammation of the lungs; short and sudden expiration after a long, slow inhalation, is a sign of congestion of the lungs; when breathing is performed with difficulty, and with a hissing noise, as if the throat were obstructed by some spongy material, and attended with a barking cough, there is inflammation of the larynx; interrupted, sighing respiration, with febrile symptoms, generally precedes some eruptive disease.

The sound and conditions of a cough are to be carefully considered. If attended by rattling, as from accumulation of phlegm in the throat, the cough is of a catarrhal character, and when hard, dry, and accompanied with cries of pain, it denotes inflammation of the lungs or throat. A dry, hollow, spasmodic cough, without apparent pain, and attended by vomiting, is peculiar to the whooping cough; a husky, crowing, metallic-sounding cough is attendant on croup.

Deviations from the natural state of the urinary and alvine excretions are to be regarded as significant symptoms; but as all such conditions have been referred to under specified disorders, they need not be repeated.

A dry heat, with rapid pulse, denotes fever. Eructations, vomiting, and an offensive breath, indicate derangement of the stomach. Continued convulsions, spasmodic movements of the limbs and muscles of the face, are, if not induced by teething, common indications of disease of the brain. Unquiet sleep, general uneasiness, irregular appetite, are manifestations of derangement, demanding attention; and no unusual appearance of any description is to be disregarded ir forming a diagnosis of infantile disease.

COLIC.

Little children appear to suffer frequently from pain in the stomach or abdomen, more especially during the first six months of life. These pains are mani- fested by violent screaming, forcible flexion of the lower limbs upon the abdomen, struggling, and gen- eral restlessness. The very reprehensible practice of giving laxatives, carminatives, chamomile tea, as well as other food than that afforded by nature, often occasions, doubtless, this seemingly distressing affection. And further, there is no more reason for supposing that the delicate stomach of the infant cannot be over- loaded, even with its most appropriate food, than that gluttony can be practised with impunity when the powers of digestion are more mature. No remedy will produce more than temporary alleviation, except under a judicious regulation of the child's diet. The medicine best adapted to relieve these colic pains, if they are not directly brought on by the use of chamo- mile tea, is

CHAMOMILLA.

The second in importance, more especially if the colic should be accompanied by vomiting or diar rhœa, is

PULSATILLA.

Should there be constipation, the remedy is

NUX VOMICA.

DOSE. — Three or four globules of either of the above may be given, in a dry state, as frequently as once every half hour, until relief is afforded.

SLEEPLESSNESS.

When a child, at its accustomed hour of rest, shows no disposition to sleep, but is uneasy and fretful, it must be regarded as suffering from some derangement of health. Under such circumstances, it is often thought necessary, by those who have charge of the child, to administer some preparation of opium as a soporific, seemingly unaware that stupefaction is not healthy sleep, and that the opiate which is producing an entirely unnatural condition for the time, and masking instead of counteracting the disturbing cause, is also laying the foundation for much future suffering.

When no cause can be detected to which sleepless ness may be attributed, the medicine to be given is

COFFEA.

The above is recommended on the supposition that, in the case of an unweaned child, the nurse does not make use of coffee as a beverage. If it is used, the medicine should be

CHAMOMILLA.

When improper diet, causing derangement of the

stomach, has produced sleeplessness, the remedies are

PULSATILLA,

If diarrhœa, or

CHAMOMILLA,

If constipation be present.

DOSE. — Three globules, to be repeated in an hour, if necessary.

CATARRH.

Infants are frequently troubled with what is commonly called " cold in the head," which may be considered a chronic catarrh, it being usually quite obstinate, and prolonged, impeding breathing through the nose at times, to such a degree as seriouly to interfere with sleep and feeding. The principal remedy for this difficulty is

NUX VOMICA.

Should this be inefficacious, the next medicine to be given is

SAMBUCUS.

DOSE. — Three globules, every evening.

MILK CRUST.

This obstinate affection of the skin is described under the head of " Crusta Lactea," (page 165,) in the chapter on Cutaneous Diseases, to which the reader will please refer.

CONVULSIONS.

(See page 125.)

DIFFICULTIES OF DENTITION.

(See pages 45–48.)

VOMITING.

When this amounts to nothing more than a regurgitation of milk, as is frequently the case, it will not be necessary to administer any medicine, as too much food taken at one time is the cause; but when all the nutriment received appears to be rejected, the medicine best adapted to restore the stomach to a healthy condition is

IPECACUANHA.

Should not this relieve, after three doses have been given, at intervals of four hours, *Pulsatilla* may be tried in the same way, and afterwards, if necessary, *Nux Vomica.*

DOSE. — Three globules, given in a dry state.

INFLAMMATION OF THE EYES.

(See " *Ophthalmia,*" Chap. VI., pages 176, 177.)

GUM OR TOOTH RASH.

This is a cutaneous eruption, occurring during infancy, and consisting of minute pimples that appear on the face, shoulders, and arms. These pimples are small and numerous, of a red color usually, when they are known as "red gum;" and sometimes are white, commonly called "white gum." They are both of the same nature, differing only in color, and generally make their appearance on the skin previous to the first teething. The medicine to be given for this eruption is

RHUS TOXICODENDRON.

Three globules, morning and evening, for three days.

If no improvement is manifest, give, in the same dose and manner,

SULPHUR.

WORMS.

(See article under this head, page 79.)

THRUSH.

(See "*Aphtha*," page 51.

INFANTILE DIARRHŒA.

An irritation of the bowels, giving rise to diarrhœa, frequently prevails with children, particularly during dentition. Indigestible food and exposure to cold are common causes, as well as the imprudent administration of laxative medicines. The disease is attended with irritability, peevishness, restlessness, want of appetite, and frequent fits of crying.

In the largest number of cases, the remedy that will counteract the cause of irritation is

CHAMOMILLA.

If, in two days, there should be no change for the better, the next medicine to be given is

SULPHUR.

DOSE. — Eight globules are to be dissolved in a wineglass full of water, and a tea-spoonful of the solution given every four hours.

CHOLERA INFANTUM.

This disorder almost always takes place during the period of dentition, from the fourth to the twentieth month of infancy, and consists of continued vomiting and purging, which causes great debility and emaciation. Pain in the stomach and bowels is manifested from the commencement of illness; the extremities are cold, the head hot, the eyes dull; and if the disease continues unchecked, convulsions and fever set in, with aphthous ulcerations in the mouth, difficulty of breathing, and spasms. This is a very frequent and dangerous complaint, occurring chiefly during warm weather, and most frequently among children who are confined in small, damp, ill-ventilated dwellings. At the beginning of the disease, should symptoms of fever exist, a few doses of *Aconite* must be given, and afterwards

1. CHAMOMILLA; 2. IPECACUANHA;

In alternation.

If the symptoms are violent at first, or if the disease progresses, the remedy is

VERATRUM.

If dysenteric symptoms are manifest at any time, give

MERCURIUS COR.

DOSE. — Ten or twelve globules of the medicine should be dissolved in three table-spoonfuls of water, and a tea-spoonful of the solution given every third hour.

DIET. — If the child is not weaned, the nurse should carefully abstain from the use of unripe fruit, from flatulent or uncooked vegetables, from any food or avoidable influence that causes disturbance to her own health. A change of air from a small, confined room to a larger and better ventilated place, or from the city to the country, has often proved of marked and immediate benefit.

19

CHAPTER IX.

In this class are comprehended menstrual irregularities, known to physicians by the terms "amenorrhœa," "dismenorrhœa," "menorrhagia," "leucorrhœa," and "chlorosis;" and the prevalence of all these varieties is attributable to the modern artificial and luxurious manner of living. They were of comparatively rare occurrence in former ages, when the habits of life were more simple and natural, and they are known to have increased correspondingly with the progress of civilization, as enervating customs, induced by wealth and luxury, have marked its march to the present time.

AMENORRHŒA,

Or obstructed menstruation, may be partial or complete, and in both cases is attended with lassitude, dyspeptic symptoms, irregularity of appetite, sometimes very capricious, occasionally with an obstinate cough, and symptoms like those of pulmonary consumption. Among its chief causes, in addition to those that in general bring on an unhealthy condition, are exposure to cold and dampness, and powerful or protracted mental emotion.

The chief remedies are, —

1. PULSATILLA,

which may be followed by

2. SEPIA;

and afterwards by

3. SULPHUR,

should the preceding prove ineffectual.

DOSE. — Dissolve eight or ten globules in a wine glass full of water, and take a tea-spoonful of the solution every fourth or fifth hour, until an improvement is observed, not more than six doses of any one medicine being taken.

DISMENORRHŒA,

Or difficult menstruation, is attended with extremely violent pain in the back and loins, and such as are described to be not unlike the bearing down pains of labor. The medicines to be used are the same as those above enumerated, with the addition, if required, of

CHAMOMILLA,

in similar doses.

MENORRHAGIA,

Or excessive menstruation, is attended with an actual hemorrhage from the menstrual vessels, and is distinguished by two varieties, in one of which menstruation is excessive at the accustomed period, and in the other, of too frequent recurrence.

The remedies are, —

1. IPECACUANHA; 2. NUX VOMICA; 3. CHAMOMILLA.

DOSE. — As above directed.

LEUCORRHŒA,

Or " Fluor Albus," affects, in a greater or less degree, nearly all who are subject to disordered menstruation, and is an abnormal secretion, dependent, in most instances, on causes affecting the general health of the system. Individual physical conditions, with varieties in the character of the secretion itself, indicate different remedies ; but by far the largest number of cases are benefited, as is true in relation to all menstrual irregularities, by the use of

1. PULSATILLA.

Other medicines that have proved the next in' usefulness are, —

2. CALCAREA; 3. SEPIA; 4. SULPHUR.

DOSE. — Four or five globules, morning and evening.

CHLOROSIS.

The morbid state of the system to which this term is applied results from some imperfection of menstrual action occurring at the period of puberty, and is attended with very great general debility, a greenish-white hue of the skin, acidity of the stomach, creating an appetite for lime and alkalies generally, with various dyspeptic symptoms. It is a complaint requiring often a long course of treatment, under the superintendence of an experienced practitioner. But the medicines that have been found the most frequently useful under ordinary circumstances, are, —

1. PULSATILLA; 2. SULPHUR; 3. BRYONIA.

DOSE. — Four or five globules, every evening,

It is not our purpose to enter into any particular description of the derangements of pregnancy. The condition itself is not one of *disease*, and therefore cannot properly be introduced here. Neither are the disturbances peculiar to it such as often need medi· cation. With the exception of certain accidental difficulties, the serious nature of which demands prompt personal assistance, which, however, can be most appropriately afforded by one of the same sex, there are no indispositions accompanying the state of preg-

nancy that may not be modified or entirely prevented by strict attention to the following hygienic
precepts.

All violent muscular exertion must be abstained
from. The gentle exercise of walking in the open
air must be continued, during the term of gestation,
and when impracticable, should be taken within the
house. An erect posture should be preserved as
much as possible. Daily bathing in water of the
temperature of the body, or colder, if well borne,
must be practised. All drugs should be avoided, and
all strong passions or depressing emotions. The diet
must be simple, yet nourishing; spices, liquid stimulants, pastry, and indigestible food of every description abstained from, together with coffee and tea,
unless either of the two last named is very urgently
desired. Animal food is not required more than
once a day. Regularity in eating, in sleeping, in
exercise, in all the habits and requirements of life,
is of great importance; and the duty of studying,
and adhering to, all the laws of health devolves with
increased responsibility on one in the situation here
supposed.

CHAPTER X.

In this chapter it is proposed to give a concise account of the nature and management of such external injuries as are liable to happen under all circumstances; and it is hoped that the description may be sufficiently explicit to enable any individual to administer prompt and efficient relief in cases of suffering from unexpected causes, when surgical aid is unattainable.

BURNS AND SCALDS.

These terms are well known as applied to external injuries occasioned by great heat, and the distinction made between them has reference merely to the nature of the agent by which the heat is conveyed. Thus, if boiling water or other heated fluid comes in contact with the skin, the injury caused is properly termed a " scald." This is usually less serious, because more superficial, than a " burn," resulting from contact with a heated substance, since boiling water, by which

(223)

a scald is produced, contains heat less intense and durable in its action. When, however, the fluid is boiling tar, oil, or varnish, the scald is more severe, and the injury assumes the character of the worst description of burns, namely, that in which the " cutis," or true skin, is destroyed.

In a practical point of view, no distinction need be made between the *burn* and the *scald*, as the treatment is, in all respects, the same.

The most simple form of this injury is that in which the skin exhibits a mere redness. A more severe burn or scald causes the skin to exude a fluid (serum), the presence of which raises up the outer skin, or " cuticle," producing vesications or blisters. The severest form of all is where the vitality of the integuments is destroyed, and mortification takes place. In the second, if vesication should be extensive, and always in the latter case, the constitutional disturbance is considerable, the pulse being feeble and rapid, the surface pale and cold, and a great degree of chilliness present.

Scalds and burns are more or less dangerous, according to the locality of the injury, the extent of surface affected, the age and constitution of the sufferer. When a large portion of the surface has been burned or scalded, even though superficially, danger is always to be apprehended. When the trunk of the body has been burned, there is more danger than when an equal extent of surface on the face or limbs has been affected. When infants or aged persons suffer from these accidents, the consequences are more serious than at other periods of life. When the burn or scald is deep or extensive, if there occur constant chilliness, with insensibility to pain, and drowsiness, there is much reason to apprehend a fatal termination.

In relation to treatment, the application of warmth, in accordance with the homœopathic law, is the safest and surest mode of relief. The object is to restore gradually the inflamed surface to its normal condition ; and this is best effected here, as in natural disease, by employing agents that produce similar symptoms or consequences. It is improper to make use of cold water, or any cooling application whatever. In . slight cases, the relief is temporary, and when the warmth returns, the pain is more severe than at first. In severe cases, the cold application is attended with decided danger. Especially is it recommended to avoid all " pain-killing " lotions or ointments, the principal ingredient of which is *lead*, in some form, that, by absorption, will sooner or later act as a poison The chief efficacy of these " quack " preparations re- sides in the oil or lard of which they are partly composed, and the benefit arising from this may be more readily obtained in the cheaper and purer form of olive oil.

1. If the burn is slight, producing a simple redness of the skin, without vesication, the best mode of alleviating the pain is to hold the burned part near the fire, if it can be conveniently done, or to place it in

very warm water, heated to one hundred and ten degrees of Fahrenheit, until the pain subsides. Should the burned part be so situated as to render this step inconvenient or impossible, it may be bathed for a few minutes in tepid water to which one tenth part of the

TINCTURE OF ARNICA

has been added, and afterwards covered by warm lint or cotton. Heated alcohol may be also used if the pain is severe and the inflammation extensive. The administration of medicine is not generally required; but should there be much pain, or great nervous excitement, six globules of

COFFEA

. will tend to alleviate; and

ACONITE,

in the same dose, may be given, if, as in children, much alarm is manifested.

2. When the burn or scald has been so severe as to occasion the formation of blisters, the best of all applications is *cotton wadding*, applied over the whole injured surface, after the blisters have been pierced with a needle in order that the fluid contents may escape. Care must be taken that, in doing this, the skin be preserved as entire as possible. The unglazed side of a thin layer of cotton should be applied next to the skin, moistened with olive oil, upon which may be laid three or four thicknesses of the cotton, one over the other, the whole being kept in place by a bandage of some kind. By this simple application warmth is retained, and the air effectually excluded;

and this is the only assistance art is called upon to afford. The medicines, if needed, should be the same mentioned in the preceding section.

3. When the burned part has been so much injured as to present a dark, dried appearance, like the outside of roasted meat, the true skin has been destroyed. In this case, a bread and milk or linseed meal poultice must be applied over the whole of the burned surface, and continued until the skin is detached from the flesh beneath. An artificial covering ought then to be produced by sifting wheat flour over the burned portion, and a short distance beyond its edge, until a crust is formed of the thickness of a cent. As a new skin is being produced, this covering of flour, which has served the purpose of absorbing all moisture exuded, will gradually fall off. Sometimes, however, it does not readily become detached, but remains as an irritating substance to the tender, denuded surface. In such condition, a reapplication of the poultice above mentioned will remove the irritating incrustation, and thus afford an opportunity of ascertaining if a repetition of the flour sifting be necessary.

Should symptoms of fever arise at any period in consequence of the local irritation, it will be proper to give

ACONITE

(six globules), and repeat, if necessary, at intervals of four to eight hours.

In describing the best and simplest method of treating burns and scalds, we have intentionally omitted mention of the variety of applications recommended, in order that the plan of this work be fully carried out, viz., the avoidance of the embarrassment that is caused by consulting a multiplicity of methods, which so often incapacitates one from rendering prompt and suitable assistance. As the principal object is the exclusion of all external irritants, allowing Nature to remedy the evil in her own way, it must be evident that beyond the accomplishment of this single purpose, every interference of Art is disturbing and injurious.

WOUNDS.

In the description of wounds, it is usual to refer to them under the four general terms of "contused," "incised," "punctured," and "lacerated." A *contused* wound, or bruise, is caused by a blow from a blunt instrument, or by a fall, and the parts below the surface are injured, the skin being preserved unbroken. An *incised* wound is one made by the cutting edge of a sharp instrument. A *punctured* wound is produced by the end of a pointed instrument, penetrating through the skin. And a *lacerated* wound is occasioned by the rough, blunt edge of an object that tears the skin, injuring the parts beneath. The character of all these wounds, in relation to their mildness or severity, their duration, and termination, depends on the degree of violence with which they are inflicted, the nature and condition of the object that produced them, and the age and constitution of the wounded person.

1. The *contused wound* is seldom serious, the unbro-

ken skin preventing external hemorrhage or bleeding.
The different hues that are observed at and around the
place of the bruise are caused by the " extravasation,"
or effusion of blood from the small vessels in or under
the skin, that are ruptured by the blow or fall. At
first there is a red appearance, which soon becomes
dark. Then it changes to a violet hue. About the
fifth or sixth day from the reception of the injury, the
parts present a green appearance ; a day or two after, a
yellow hue ; and the skin gradually resumes its natural
color, more slowly or more rapidly, according to the
health of the injured person, and the amount of ex-
travasated blood. The pain or soreness consequent on
contusion may be much relieved by the external application of

ARNICA,

one part of the " Tincture " to six parts of water. This
application may be made three times daily, for five or
ten minutes, and six globules of *Arnica* may be taken
internally at the same time, should much pain or soreness be present.

2. The *incised wound*, a clean, and generally straight
cut through the skin, is always attended with immediate hemorrhage from the severed blood-vessels, and
if the flow of blood be profuse, the first thing to be
done is to check the bleeding. When an artery is
wounded, the blood suddenly springs out in jets, cor-
responding to the beating of the pulse, and is of a bright
scarlet color. When from a vein, the blood flows
steadily, and is of a purple hue. The bleeding from

20

veins or very small arteries may be arrested by pressing
the sides of the wound firmly together, after washing
away all particles of dirt or other substance, and retained in position by strips of adhesive plaster, that
made from *Arnica* being the best. The plaster should
be cut in narrow strips, wider at the ends than in the
centre, that apertures may exist just above the incision
for the escape of any fluid, and of a length to extend
an inch or more beyond each side of the wound. These
strips, after being warmed sufficiently to adhere, should
be applied across the cut, — not lengthwise, — and
carefully laid down, so as to bring the edges of the
wound in close contact. A portion of lint or folded
linen cloth, of a size sufficient to cover the injured part,
is then to be applied, and retained in place by a bandage extending around the limb or body. When the
wound is closed in this manner, an adhesive secretion
issues from the joined sides, which, at the end of a
week, or shorter time, becomes so hard and tenacious
as to hold the sides together without external aid, and
a communication is reëstablished between the opposite
surfaces. This process of Nature's healing is called
" union by the first intention." In slight, simple·
incisions, nothing more is required than thus to bring
and retain the edges in contact. The presence of any
oil, balsam, or ointment in the wound is a constant
source of irritation, and a positive impediment to a
healthy adhesion. Such " all-healing " applications are
very much worse than useless. If the sides cannot be
made to adhere in consequence of the flow of blood, the

use of very cold water, either in repeated bathing, by means of a sponge, or by wet cloths, is the best of all methods to arrest the hemorrhage.

If a large artery has been wounded, which may be known by the forcible expulsion of bright red blood in jets, the course above mentioned will not be sufficient. Under such circumstances, presuming that an upper or a lower limb has been injured, it will be necessary to find the artery above the wound, which may be recognized by its pulsation, and to press upon it with the finger ; or if the artery cannot readily be found, to tie, in a knot around the limb, a handkerchief or other bandage between the wound and the heart, and then to pass a small, strong stick under the knot, and to turn it until the twisted handkerchief encircles the limb so tightly that the flow of blood is checked. This pressure must be continued, since it can only prove a temporary expedient, till the assistance of a surgeon can be obtained, who will immediately proceed to secure the severed end of the artery. In case of the impossibility of procuring such aid within four hours, as the circulation ought not to be, for a longer time, obstructed by the bandage, a bystander may perform the operation of tying the artery, in the following manner : Prepare a ligature, by waxing three silk threads, a foot in length, twisting them together ; and also procure a small hook, or pair of pincers. Then, after washing the wound with warm water, request an assistant to loosen slightly the bandage, which is supposed to have been, up to this time, tightly twisted, and the end of the bleeding vessel will be seen throwing out blood at every pulsation. Draw this vessel out by the pair of pincers, so that the ligature may be passed around it and tied with a double knot, lightly, but sufficiently firm to arrest the bleeding. Care must be taken to include within the ligature no other than the bleeding vessel. If more than one artery has been severed, it will be necessary to secure all in the same manner. The edges of the wound should then be brought together and retained by adhesive straps and bandage, as above directed, the ends of the ligature being cut off.

It will be perceived that the object in dressing incised wounds is to retain the divided surfaces in close contact until union takes place. This is not always practicable by the use of plasters or bandages, especially in wounds of the ear. lips, nose, or throat. In such cases, stitches made by a needle and thread, or " sutures," as they are called, will be required. As these sutures are liable to excite inflammation, they are not resorted to when they can possibly be dispensed with. A needle is to be threaded with waxed silk doubled, and the point of the needle passed through the skin a short distance from the edge of the incision, and carried through the opposite side at the same distance. The thread is then to be cut near the needle, and the two loose ends drawn together by tying, so as to bring the opposite surfaces of the wound in contact. If other stitches are required, they should be made in the same manner, leaving the ends of each untied till all the necessary stitches are taken, when double knots in all are to be made.   When permanent union is effected, the threads are to be divided, and gently withdrawn from the wound. As before remarked, this operation is never performed now excepting under circumstances of unavoidable necessity, as it is often productive of considerable pain and irritation. In concluding this description of the different methods of treating incised wounds, it is important to remark that the position of the wounded part should be such as to relax the muscles as much as possible, that the divided surfaces may not be drawn asunder ; and this position is to be maintained whenever practicable, until firm adhesion has taken place.

For the debility consequent on loss of blood, the remedy is

CINCHONA,

eight or ten globules of which may be given at one time, and repeated if required. Should there be much local irritation or pain,

ARNICA

is to be given in a similar dose ; or should feverish symptoms arise,

ACONITE

is to be administered in the same manner: and the two last named remedies are to be alternated every four to six hours, so long as the above symptoms exist.

3. The *punctured wound*, made by the sharp point of any instrument or object, is an injury of small extent externally, greatly disproportioned, in general, to its depth, and more dangerous than an incised wound, though attended with much less hemorrhage. If the

20 *

object causing the injury remains in the woun l, the
first step, of course, is to extract it. This cannot
always be done without enlarging to some extent the
punctured place, though it should never be attempted
unless really necessary for the purpose of extra:tion.
Th3 punctured wound is liable to be followed by inflammation, and this must be, as much as possible,
counteracted by cold applications — the best method
known for such a purpose. Two or three thicknesses
of cloth, kept constantly wet with cold water, :nav
check inflammation, and allow the wound to heal by
union, as in simple incisions ; but should there be no
hope of avoiding suppuration, or the formation of mat·
ter, — the result of a high degree of inflammation, - - a
bread and milk poultice must be kept applied, and
daily renewed until inflammation subsides, or the dis
charge of matter ceases.

1. ACONITE; 2. ARNICA;
should be given in alternate doses, as recommended ʼ
the preceding section.

4. The *lacerated wound*, where portions of the skin
and parts beneath are detached or torn apart, is at
tended with less hemorrhage generally, but is more
serious than an incised wound. It is subject to more
severe inflammation, and causes greater constitutional
disturbance. The first thing to be done is to remove
all dirt and other substance from the wound, by wash-
ʼing it carefully with warm water, and then, if no large
artery has been wounded, to replace the lacerated portions of the skin in their natural position, bringing

them in contact by strips of plaster, if necessary, that they may unite by adhesion. The whole is then to be covered by cloths wet with cold water, to obviate that degree of inflammation which would prevent union by the first intention. If the laceration be extensive, and portions of skin entirely detached, so that contact cannot be effected, the best application is a warm bread and milk poultice, over which a small quantity of sweet oil is spread. The poultice ought to be renewed daily, or twice a day during warm weather. For all kinds of wounds, cold dressings are to be used, in order to reduce inflammation, thereby promoting union; but when from the violence of the injury, or other cause, inflammation seems to be terminating in suppuration, warm applications will hasten the process and the cure, which cold, under such circumstances, would retard. The chief remedy to be administered for the promotion of suppuration is

HEPAR SULPHURIS,

which is to be given in doses of six globules, and repeated on every renewal of the poultice. *Calendula* has been of late recommended as an external application in the treatment of incised and lacerated wounds, while *Arnica* has been preferred as a remedy for injuries when the skin remains unbroken. It is to be mixed with water, in the proportion of one part of the tincture to ten parts of water.

Wounds made by a *musket* or *pistol shot*, or any small substance so forcibly thrown as to penetrate through the skin, partake of the character both of punctured and lacerated wounds, and should be treated as such, after the extraction of the substance by which the injury was caused. If the object has not penetrated deeply, and can be felt near the surface, it may be removed by the finger; but it should be understood that there is to be very little probing, since it is better that the object or objects remain in the wound than that the already existing inflammation be increased by awkward attempts at extraction. A bread and milk poultice is often sufficient to bring to the surface any slightly embedded substance. If nothing remains in the wound, and the bleeding has ceased, a piece of lint dipped in cold water should be laid over the punctured spot, and secured by strips of arnica plaster. At the same time, six globules of

ARNICA

must be given, and the dose repeated twice a day while there is local pain or soreness. In case of the occurrence of feverish symptoms, a similar dose of

ACONITE

should be alternated with

ARNICA

every fourth hour.

DIET AND REGIMEN. — For all injuries of the nature above described, the food should be simple and digestible, stimulating articles of every description strictly avoided, a state of entire rest in a moderately-cool apartment enjoined, and cool beverages taken — the best being cold water. Cold local ap-

plications must not be made too extensively, particu
larly if the injury is severe, and situated on the body,
or if the injured person is enfeebled by loss of blood,
or if young, or very aged.

SPRAINS.

The word "sprain" is used to denote a sudden and
forcible extension or stretching of the tendons that
pass over joints, or of the ligaments that connect the
bones together, with, in severe sprains, a rupture of
their fibres. Violent pain is immediately felt after an
injury of this kind, attended with inflammation and
swelling, succeeded by a weakness and stiffness of the
joint. The wrist and ankle joint are the most exposed to this injury, though it may occur in the knee,
and other large joints. If the sprained limb be not
kept entirely still, or if a large joint be affected, there
will result considerable pain and fever, which, though
but seldom endangering life, will subject the sufferer
to a long and tedious confinement.

The treatment of a sprain consists in the external
application of diluted

TINCTURE OF ARNICA,

together with the internal administration of the same
in globules, six at a dose; after which, if pain continues, the diluted

TINCTURE OF RHUS TOXICODENDRON

may be substituted for the *Arnica*, and a dose of the
same taken in a like manner. The water with which

the tincture is mixed should be cold, and used three
or four times daily; and during the intervals, cloths
wet with cold water are to be laid on the injured
place. Every morning and evening, cold water may
be poured from a pitcher upon the sprained part.
This last application is also beneficial for weak joints.

DISLOCATIONS.

It would not be in agreement with our proposed
plan of condensation, neither would it comport with
the character of a domestic medical treatise, to enter
into a full description of all the dislocations that are
liable to take place. Many of them would not be
easily distinguished or replaced by one ignorant of
the construction of the different joints, the disposi
tion of the ligaments that connect them, &c.; and
without a knowledge of anatomy, no person is properly qualified to manage dislocations in general. But
there do occur cases of displacement that are comparatively easy of treatment, and which, if neglected,
would be attended with a great amount of needless
inconvenience and suffering.

When a dislocation occurs, the extremity of a bone
is thrown out of its natural position; that is, in most
cases, a projection of one bone forcibly removed from
a cavity in the adjoining or opposite bone. The ligaments uniting the bones or binding them to each other
are lacerated, when the bones leave their proper relative position. The muscles also connected with the

disi cated bone are elongated; and if the bone is not soon made by external force to resume its natural place, it becomes strongly held in its new position by an adhesion of the ruptured ligaments to parts with which they come in contact, and by a contraction of the elongated muscles. The longer the bone remains displaced, the greater will be the difficulty of returning it. Hence the necessity of prompt, intelligent assistance.

There are dislocations of the jaw, shoulder, elbow, wrist, and fingers; of the hip joint, the knee, the ankle, and the toes; of the neck, the spine, and the collar bone; and dislocations in combination with fractures. Descriptions of such only as may be readily replaced by an unprofessional person will be given, for the reasons above mentioned: and first, of

DISLOCATION OF THE JAW.

The under jaw is sometimes displaced from the upper by blows on the chin, and by the act of yawning. The end of the lower jaw slips from its socket in the upper, generally on both sides, and remains immovable, with the mouth wide open, and the chin projecting. Were not such a position obviously and painfully constrained, the expression manifested would be extremely ludicrous. The reduction of the dislocation may be accomplished in the following simple manner: The operator's thumb of each hand should be placed as far back as possible between the jaws, and the fingers be clasped together under the chin, so that the latter may be drawn upwards while the thumbs are pressing down the back part of the lower jaw. This downward pressure by the thumbs, together with the elevation of the chin, will replace the joint in its proper position. In doing this, the operator must stand behind the patient, taking care to slip his thumbs down between the jaw and cheek of either side when the bone is felt slipping into its place, as the meeting of the teeth together is sudden and forcible.

Another mode consists in placing two pieces of cork between the teeth, as far back as possible, to act as a fulcrum while the chin is being elevated; or two small pieces of hard wood may be placed in each corner of the mouth between the back teeth, and held by an assistant, while the chin is raised by passing both hands under it. The elevation of the chin should be even, that both sides of the lower jaw may be acted upon equally; otherwise only one side of the jaw may be reduced, and again dislocated during subsequent attempts to replace the other. When the joints are restored to their places, the mouth should be kept closed for a few days, by a handkerchief or bandage passed twice round the head and under the chin, and all food taken in a liquid form, so that mastication may not be required. Should there be much pain afterwards, every part uncovered by the bandage under the ear and upon the cheek may be rubbed with

ARNICA,

one part of the tincture to ten parts of water.

DISLOCATION OF THE SHOULDER.

This dislocation is more liable to happen than any other, as the arm is always extended to lessen the force of a fall, and the socket in the shoulder bone on which the arm bone turns is not a deep one. There are four directions in which the arm may be thrown from its socket; but the one most frequent, and the most easily managed, is where the head of the bone falls into the armpit. This dislocation may be known by the impossibility of bringing the extended elbow to the side, by a round prominence in the armpit, and a depression on the top of the shoulder. The reduction is to be accomplished by placing one hand on the prominent point of the shoulder just above the cavity made by the absent bone, grasping the arm above the elbow with the other hand, and pulling gradually and firmly outwards from the body. If this attempt prove unsuccessful, the injured person should lie upon the floor; the operator, being seated at his side, must place his foot, without shoe or boot, in the patient's armpit, grasp the dislocated arm at the wrist, and pull obliquely downward and outward, until the arm yields to this extension, when it may be gently forced to approach the side, extension being continued, so that the head of the bone will glide over the edge, and into the socket. Sometimes this is readily effected; at other times great strength and adroitness are required. After the reduction, the arm must be confined to the side, with the hand in a sling, for a fortnight at least,

21

that the ruptured ligaments may have time to reunite. The local pain may be relieved by bathing the shoulder with diluted

ARNICA,

as directed in the foregoing section.

DISLOCATION OF THE WRIST.

This is a displacement of the bones forming the wrist from the two bones of the forearm; and it may be known by the unnatural position of the hand, it being thrown backward or forward, when the displacement is complete, or from both bones of the arm, and twisted, if partial, or from one bone. To reduce this dislocation, the injured arm should be seized above the wrist, and held firmly by an assistant, while the hand is grasped and steadily extended, until there is sufficient yielding to the extending force to allow the protuberant bones to be pushed back into their natural position.

ARNICA

may be applied, as for all other dislocations, and the hand kept entirely still for several days.

DISLOCATION OF THE KNEE.

A dislocation of this kind is too serious to be submitted to domestic treatment, but occasionally the small bone that protects the knee joint in front, called the "patella," or kneepan, may be displaced; and when this accident happens, it is attended with violen

pain, and the displacement may be easily seen. To
return the bone to its place, the limb should be extended at a right angle with the body, and in this
position the patella may be easily moved. The knee
must be kept extended and at rest for a week or
more, and

ARNICA

applied if pain and inflammation exist.

The reduction of dislocations other than those above
described demands that degree of skill and anatomical
knowledge which is only to be obtained by much study
and practice. Beyond the statement that, in all cases,
the first step is to apply an extending force to the displaced limbs sufficient to bring the parts in coaptation,
it is not advisable to proceed here, since the more
serious dislocations, especially if complicated, as they
frequently are, with fracture, might be rendered by
misunderstanding and mismanagement altogether irre
ducible.

FRACTURES.

The same remarks that have been made in reference
to dislocations will apply to fractures, viz., that in the
larger number of such injuries no one except a professional man can be safely trusted with their management. Nevertheless much general information may be
afforded which can be made available, in the unavoidable absence of a surgeon, to the promotion of the
sufferer's comfort, and the avoidance of a life-long de

formity. For the judicious, limited information, under the circumstances referred to, is better than none; and, as previously intimated, the services of the medical man, whenever possibly procurable, are not to be and cannot be superseded by one possessed of the partial knowledge here afforded. All that is to be written on the subject is intended for those only who are so unfortunate as to be beyond the reach of professional attendance.

Fractures are usually occasioned by a fall, and may be recognized by pain at the place of the injury, by swelling, deformity, and loss of voluntary motion, with a crackling or grating sound when the parts are moved in opposite directions by the hand. They are termed " transverse," when the bone is fractured horizontally; "longitudinal," when it is broken lengthwise; " oblique," when the line of fracture is at any angle between the two. When the fracture is not connected with a wound externally, it is termed " simple; " when thus connected, " compound ; " when the bone is broken in several pieces, it is termed " comminuted ; " and when dislocation occurs with it, " complicated."

Whenever a bone is fractured, the injured parts should be placed in a position as easy and natural as possible until proper attention can be procured. Thus, if one of the lower limbs is broken, and the accident occurs from home, — in the street, for example, — the easiest and readiest mode of conveyance is upon a window blind or shutter, which may be covered with blankets, straw, or clothing of any kind. The injured person should be lifted upon this by four assistants, the broken limb laid straight by the side of the other, and fastened to it by a handkerchief or other bandage. The bearers should walk carefully, keeping step with each other, and with as little jolting as possible, since every irregularity of motion causes severe pain and tends to widen and even complicate the fracture.

If the arm be broken, it should not be allowed to hang down, but be placed in a sling formed by a handkerchief, the base being made broad enough to support the forearm from the elbow to the fingers; and, thus conveyed, the injured limb would be less subject to motion than in any other position.

FRACTURE OF THE RIBS.

This injury is occasioned by a fall or a blow, and may be recognized by a sharp pain felt at every inspiration in the spot where the injury was received. If the hand be placed on this part of the chest or side, . and the person be requested to cough, the ends of the broken rib will be felt moving on each other, and a grating sound perceived. The treatment consists in confining the ribs so that they may not move during respiration. For this purpose, a linen or flannel bandage, eight or ten inches wide, and five or six feet long, is to be tightly wound about the chest, and secured by sewing. This should be worn for a month. If the injury be very severe, and there is a fracture of both sides, or of the breast bone, the bandage must not be applied.

21 *

ARNICA

(six globules) ought to be administered, if there is much pain ; and if feverish symptoms are perceived,

ACONITE,

in the same dose, may be given, and repeated twice daily so long as cough or fever exists.

The food should be simple and unirritating, and the injured person be kept as motionless as possible.

FRACTURE OF THE CLAVICLE.

When the clavicle, or collar bone, is fractured, the accident may be known by a roughness or irregularity being felt at the injured part, with swelling, and a falling of the shoulder towards the side, carrying the arm forward on the front of the chest. The reduction is effected by drawing the shoulder back to its natural position, and on a level with the one opposite. That it may be retained in this position long enough for the bones to unite, a wedge-shaped, unyielding substance, as pine wood, must be made, nearly two inches thick at its upper part, tapering to an edge and reaching to the elbow, covered with several thicknesses of cloth to protect the skin ; the broad end of this wedge should be placed under the armpit, the arm brought down to the side, and there held by a bandage passed over the arm of the injured side and around the body. The forearm is to be supported in a sling. This position must be retained for a month at least. One or two doses of

ARNICA

may be given to lessen pain and prevent inflammation.

FRACTURE OF THE ARM.

This fracture is to be recognized by pain, an unnatural motion of the bone, the impossibility of voluntarily raising the elbow, and a grating sound made by the fractured ends of the bone when moved on each other. It is to be replaced by grasping the wrist while the shoulder is held firmly by an assistant, and steadily extending the arm till in appearance and length it resembles the opposite one. Four "splints," or pieces of smooth, thin wood, or thick pasteboard, previously prepared, two inches wide, and of a length sufficient to extend from the shoulder to the elbow, well lined with some soft material, as tow or cotton, are to be placed on the sides, front, and back of the arm, and there retained by a bandage, the forearm resting in a sling. This unyielding external application is necessary to prevent any subsequent displacement of the ends of the bone, and to support them in a straight line till reunion takes place, which will be in four or five weeks.

In fractures of the forearm, composed of two bones reaching from the elbow to the wrist, the same treatment with splints is required, that the bones may be kept straight until reunion is established, after the reduction has been effected by extension made from the wrist, and counter-extension by holding the arm firmly above the elbow. When the splints are applied and secured by a bandage, the forearm is to be supported in a sling for a month or more.

ARNICA

may be used as above directed.

FRACTURE OF THE WRIST AND FINGERS.

A fracture of the wrist is a rare injury, and requires for its production a very forcible blow or heavy fall, which would be likely to produce, in addition, an external wound. After any projecting bone has been pushed back into its place, a broad splint is to be applied to the front of the wrist and hand, extending to the ends of the fingers, and well padded, that uniform pressure may be maintained. A bandage should be passed several times over the splints, and the hand be supported in a sling.

·If the bones of the hand are fractured, a splint is to be applied in front of the wrist and hand in the same manner, of length and breadth sufficient to cover the lower half of the arm, the wrist, and the fingers. And if the fingers are broken, four narrow splints are to be applied, bound down with a piece of tape, after the fractured ends have been placed in apposition by the proper extension.

ARNICA

may be used, as heretofore recommended.

FRACTURE OF THE LOWER LIMBS.

The bone above the knee, the thigh bone, and the bones of the leg, from the knee to the ankle, may be broken in several places ; and these fractures are the most difficult of management, on account of the strength of muscle to be overcome in the reduction, the long confinement and the absolute rest required for permanent union. Many mechanical contrivances have been adopted in the treatment of such fractures, for the manufacture of which a carpenter's services are indispensable. The best of these is a frame in which the limb is fastened, and gradual extension made by means of a screw. But as all apparatus of this kind requires time for construction and more mechanical skill than is always available, no further reference will be made to them ; but the simplest method will be described that is known for treating a fractured thigh bone without splints or machinery of any kind.

When such a fracture occurs, the injured person must be laid upon a mattress, under which a board is placed that will not yield to the weight of the body, of the width and breadth of the mattress. The body should lie in a straight position, with the limbs close together. The limb being held firmly at the hip by both hands of an assistant, the operator is to grasp the limb above the ankle, and draw it gradually down, by gentle, steady extension, without disturbing the straight position, until the knee of the fractured limb is brought to a level with that of the uninjured one, and a handkerchief or bandage passed around both limbs, just below the knees, sufficiently tight to prevent one knee slipping from the other. The ankles of both limbs are then to be tied together in the same manner. The feet should also be bound to each other. A pad, an inch thick, composed of cotton or tow, is to be placed between the knees, one between the ankles, and another between the feet, that uncomfortable pressure by close contact may be obviated. If by this simple method extension of the injured limb can be preserved until a firm union of the bone is effected, it will be all that the most complicated apparatus can accomplish. This is not always, even in simple fractures, to be expected, as the muscular contraction is sometimes too powerful to be counteracted by such means; but it will serve, at all events, as a good temporary expedient till professional aid can be procured.

The two bones that extend from the knee to the ankle may be fractured by the same cause of injury, or one may be broken while the other remains uninjured. In the latter case the uninjured bone will serve as a splint for the broken one, and no extension or external support will be required. Rest only is necessary.

If both bones, however, are fractured, — an accident known by pain, unnatural motion, and an unusual prominence at the place of injury, — the limb should rest straight by the side of the other, and two splints placed, one on the outside, the other on the inside of the limb from the knee to the ankle, and kept in place by bandages; care being taken that both limbs remain of an equal length, the foot of the injured limb preserving its natural position. The splints must be well padded, that injurious pressure may not be any where exerted, and two circular holes cut in their lower ends to receive the projecting ankle bones. The limb must be kept at rest for a month.

ARNICA

may be used for this, as for all other fractures and dislocations. And

ACONITE

also, in the event of any accompanying fever.

Fractured bones of the foot are to be managed like those of the hand and fingers.

ASPHYXIA. (*Suspended Animation.*)

The term " asphyxia " is used to denote an interruption or temporary suspension of the vital powers by any cause that obstructs respiration. This state may as well be expressed by the word " suffocation." It is produced when, in any manner, air is prevented from entering the lungs, whether by any mechanical obstruction of the larynx or air passage, either from within or without; by submersion in water or other liquid ; by the inhalation of gases that do not furnish oxygen ; by complete closure of the mouth and nostrils. The chemical alteration which the blood undergoes in the lungs by contact with the oxygen contained in the atmosphere is prevented by these various obstructions. Life cannot be sustained without the circulation of blood that has been " oxygenated " in the lungs. And when death occurs from suffocation, it is owing to the venous blood that is brought to the lungs being deprived of this necessary oxygenation, and being forced to pass unchanged from the lungs to the heart, the action of which latter organ it is wholly inadequate to support or excite. Even, however, after the heart has ceased to contract, should air be readmitted to the lungs by artificial respiration, the venous blood remaining there will sometimes become revivified or " arterialized," recommence to circulate by gradually increasing progress, and life be restored. Hence the great importance of unremitting exertions to reanimate those who are apparently dead by suffocation, from whatever cause induced.

SUSPENDED ANIMATION FROM DROWNING.

In the event of drowning, the body, on being taken from the water, must be at once conveyed to a house, with the head and shoulders raised, and put into a warm bed. The mouth and nostrils should be first cleansed, the wet clothing removed, the body wrapped up in blankets, and the limbs rubbed with heated cloths. Bottles filled with warm water or heated bricks may be placed under the arms and the soles of the feet. The nostrils are to be closed, and the mouth of a healthy person applied to that of the lifeless one, and the process of breathing imitated as nearly as possible; or, if this method should be ineffectual, the pipe of a bellows may be inserted into one nostril, the other being closed, as well as the mouth, and the bellows gently blown, so that the lungs may be inflated with air. The upper part of the windpipe should be gently drawn downward, and pushed backward, that the air may pass freely to the lungs, and slight pressure made upon the chest with the hand, after the lungs are inflated, that the air may be removed. This imitation of natural breathing should be continued until signs of restoration appear; and when the person is able to swallow, a spoonful of wine may be given. It is perhaps needless to add, that no time should be lost in any of the steps taken, and that efforts to reanimate be not too hastily relinquished.

The humane society established at Paris for the restoration of the drowned, stated, in their first report, that in twenty-three instances of complete restoration to life, one individual had been under the water forty-five minutes; four of the number, half an hour; and three, for fifteen minutes.

SUSPENDED ANIMATION FROM HANGING.

In apparent death from hanging or choking, all clothing ought to be removed from the throat and chest. The body must be placed in an easy position, the head and neck rather elevated, and not bent forward or backward. The body and limbs should be gently rubbed with warm cloths, and heated stones covered with flannel applied to the neck, armpits, and feet. Recovery does not so frequently take place under the circumstance of strangling as after submersion, unless restorative measures are undertaken immediately after insensibility occurs; for, superadded to the non-admission of air into the lungs, a tight cord about the neck obstructs the reflux of

22

blood from the head to the heart, thereby causing compression of the brain. If, after animation is restored, there should be feverish symptoms,

ACONITE

is to be given, as in all similar cases.

FROM FREEZING.

When a person, from long exposure to excessive cold, has become insensible, he should not be subjected to any degree of heat, or to currents of air, but be carefully removed to some sheltered place, and covered, excepting the mouth and nose, with snow, if to be procured; if not, then placed in an ice-cold bath, and kept in it for a few minutes. When the limbs become pliable, all the clothing should be removed with care, and the skin thoroughly rubbed with snow, or with cloths dipped in ice water, until the whole surface becomes red. Then the person is to be placed in a dry bed, and rubbed with cold flannel till symptoms of animation are perceived. When capable of swallowing, a tea-spoonful of strong coffee may be given, and repeated every five minutes. No external warmth should on any account be applied, and the direct heat of a fire ought to be avoided for several days. A spoonful of wine and water may be given instead of the coffee, if the person greatly desires it. Should pain be felt while recovering,

ARSENICUM

(six globules) should be administered. If the pain is more especially in the head,

ACONITE

may be given, in a similar dose.

FROM LIGHTNING.

If a person is struck down by lightning, and remains insensible, he should be placed in a current of cool, fresh air, and cold water repeatedly thrown on the face and breast.

NUX VOMICA

is the only internal remedy that promises to be of service. It should be given in doses of six or eight globules, and used as an external application at the same time, in the proportion of fifteen or twenty globules to half a pint of water. C. Hering recommends that the person struck by lightning be placed in newly-dug earth, and be covered with it, so that only the face is free, and that turned towards the sun; the position of the body half lying, half sitting. When animation returns,

NUX VOMICA

is to be given as above mentioned.

FROM HUNGER.

When a person has been deprived of sustenance for a long time, and apparent death ensues, the utmost caution is to be exercised in giving food. In any degree of positive suffering from hunger, great injury may be occasioned by a free indulgence of the appetite. When unconsciousness is the result of long deprivation of food, nothing at first should be given but a drop or two of warm milk, and the quantity may be gradually increased to a tea-spoonful. Afterwards, as consciousness returns, a little broth may be allowed, and then a few drops of wine. The amount given of any thing must be very small. After sleeping, a light meal may be allowed, but not before; and for several days after recovery, the person must partake sparingly of the most simple and digestible food.

FROM A FALL.

It was formerly a more common practice than at the present time, although not yet altogether in disuse, to plunge a lancet into the arm as soon as possible after insensibility from a fall, for the humane purpose of restoring the injured person, but with the very likely prospect of destroying every chance of recovery. Bleeding is not required, and should never be attempted. In such a case,

ARNICA

is the suitable remedy, as for all the consequences of external injury. Should the reaction be violent as consciousness returns, it may be controlled by

ACONITE.

The doses of either are five or six globules, with the external application of

ARNICA TINCTURE

to the bruised places — one part of the tincture to eight parts of water. If insensibility is produced by a fall on the head, a surgeon's attendance becomes necessary.

FROM INHALING POISONED AIR.

The carbonic acid gas found in long-closed cellars, in wells, and other places, and that which arises from the burning of charcoal or lime, will very soon produce insensibility, and death on continued exposure to it. The insensible person must be immediately removed into pure, fresh air, quickly undressed, and the head, neck, and chest be freely bathed with cold water, or, what is better, the whole body plunged into cold water; at the same time, the face may be sprinkled with vinegar and water, mixed in equal proportions, or a sponge dipped in vinegar held before the nose and mouth. The feet, hands, and arms should be well rubbed with warm flannel. If insensibility continues, artificial respiration must be promptly adopted. The process has been already described, but needs repetition. The bellows may be used, or, what is more convenient, the hands alone, in the following manner: Both hands of one individual are to be pressed upon the breast bone of the insensible person, while another pushes up at the same time the abdomen towards the chest After a short, simultaneous, and firm pressure in this manner, the hands of both individuals should be suddenly removed. The chest, freed from pressure in front and below, will expand to its usual dimensions, and air will be forced into the lungs to fill the vacuum caused by this sudden expansion. The operation should be repeated until natural breathing takes place, or until all hope of reproducing it is over.

22 *

FAINTING.

When a fainting fit occurs, the affected person should be immediately laid in a recumbent position, the head being no higher than the body; all clothing should be loosened about the neck and chest, and cold water sprinkled on the face. These means will usually suffice for restoration; but should insensibility continue,

SPIRITS OF CAMPHOR

may be applied to the nose, but not held there long.

Fainting results from various causes. When from fright, the medicine to be given is

OPIUM.

When from grief,

IGNATIA.

When from loss of blood or other debilitating cause,

CINCHONA.

When from great mental exertion,

NUX VOMICA.

When from great physical exertion,

VERATRUM.

When from violent pain,

ACONITE.

When from any strong emotion,

CHAMOMILLA or IGNATIA.

The medicines selected must be given frequently, (every five minutes;) and if after a second dose of the same no benefit is perceptible, the selection has not been a good one, and another remedy must be sought for. The free inspiration of fresh air is necessary, and therefore attendants should not be allowed to crowd around the patient. A confined, vitiated air is not unfrequently the immediate cause of fainting. It is evident, in such cases, that pure air, and plenty of it, is of the first necessity in the restoration of consciousness.

ON POISONS.

In all cases of poisoning, prompt measures should be taken, first to excite vomiting, then to administer antidotes, and afterwards to remove subsequent uncomfortable sensations by medicinal action. For the purpose of causing vomiting, warm water should be administered in large quantities; or the throat may be tickled with the feathered end of a quill; or snuff, or mustard mixed with salt, put upon the tongue. Poisoning is to be apprehended when violent symptoms suddenly appear after partaking of unaccustomed food or drink. When the nature of the poison swallowed is unknown, the whites of several eggs should be beaten up, mixed with cold water, and drank abundantly, especially if violent pain is felt in the stomach or bowels; and very strong, hot coffee, without milk or sugar, should be given, if there exists drowsiness, unconsciousness, or delirium. The antidotes for mineral acids are magnesia, chalk, or white castile soap dissolved in warm water; for alkalies, vinegar or lemon juice. Mucilaginous drinks should be freely administered in most cases of poisoning. Sugar and water is an important remedy.

PARTICULAR POISONS, WITH THEIR ANTIDOTES.

Arsenic is to be neutralized by chalk, sugar and water,* soap water, the white of eggs, or sesquioxide of iron. Any mucilaginous fluid may be freely used to promote the operation of vomiting, and to protect the mucous membrane of the throat and stomach from the corrosive action of the mineral. If symptoms of inflammation become evident, as secondary effects, with pain or tenderness of the abdomen, a few globules of

ACONITE

may be required. If nausea is present, after the severity of symptoms has subsided,

IPECACUANHA

should be given. If great irritability and restlessness,

CINCHONA.

If after this there should be nausea, vomiting, coldness of surface, with great debility, give

VERATRUM.

ACID, (Nitric.) Copious draughts of water are required, containing powdered chalk or calcined magnesia, in the proportion of an ounce of magnesia to a quart of water, or a solution of soap in warm water.

---

* Sugared warm water, in large quantities, many tumblers full, is as efficient as any antidote, and perhaps more readily procurable.

Half an ounce of soap in a quart of water, a tumbler full of either mixture being given every five minutes. No emetics, as such, are to be used.

ACONITE

should be given on the first symptoms of inflammation after free vomiting.

Acid, (Sulphuric.) The same treatment as for the above is required.

Acid, (Muriatic.) The same treatment as recommended for the two preceding acids.

Acid, (Oxalic.) The same as above.

Acid, (Prussic.) Six grains of tartar emetic is to be immediately dissolved in a large tumbler full of water, and one half of this given. If in fifteen minutes free vomiting has not taken place, the remainder should be given. Strong coffee, afterwards, in large quantities, given warm, without milk or sugar, prepared by pouring a quart of boiling water on half a pound of coffee, and straining through a cloth after ten minutes.

IPECACUANHA or NUX VOMICA

may be given for the subsequent disturbances.

Alum. Large quantities of soap suds, or sugar and water, should be drank, until vomiting is produced when

PULSATILLA or VERATRUM

can be given.

Antimony, (Tartrate of.) Any mucilaginous fluid, as gum arabic, or barley water, or slippery elm bark tea, are to be given, to promote the expulsion of the poison ; and a decoction of oak bark or Peruvian bark should be swallowed in large quantities, as an antidote to that which may remain unremoved.

ALKALIES — (as Pot Ash, Saleratus, Salts of Tartar, Pearl Ash, Smelling Salts, or Hartshorn, Soda, Lime) — are antidoted by any strong vegetable acid, as vinegar, lemon juice, &c. Large quantities of acidulated water, prepared by putting two table-spoonfuls of vinegar, or the juice of one lemon into a tumbler of water, must be given. No irritating emetics should be administered for alkaline poisoning. In case of the difficulty of procuring the acids, sweet oil will also neutralize the alkali, by converting it into soap. Any symptoms of inflammation that are manifested indicate, in all cases of poisoning, the medicine given for inflammation from other causes.

BARYTES. The different forms of this mineral, the carbonate, chloride, or nitrate, produce effects similar to those of arsenic. Alum, magnesia, lime, or soda antidote this acid by forming an insoluble and harmless compound.

COPPER, with its various salts, the blue vitriol, or sulphate; the mineral green, or hydrated oxide; verdigris, &c., which are used in confectionery and preserves, is neutralized by the white of eggs. If not procurable, use milk or flour gruel in large quantities.

CAMPHOR, as a poison, is antidoted by strong coffee, after which three globules of

OPIUM

should be given every hour.

LEAD. All the preparations of this mineral — the acetate, (sugar of lead,) the protoxide, (litharge,) the deutoxide, (red lead,) the carbonate, (white lead) — are strongly poisonous. The antidote is Epsom Salts, one dessert-spoonful, dissolved in a half pint of water, the whole drank at once. The medicines afterwards needed may be

1. BELLADONNA; 2. OPIUM; 3. NUX VOMICA.

MERCURY, and its preparations,— calomel, corrosive sublimate, red precipitate, &c., — are antidoted by albumen and gluten, the first found in the white of eggs, the last in wheat flour, in the proportion of ten eggs beaten up in two quarts of water, a tumbler full being given every two minutes. Either one of these should be given freely in sugared water, or mixed in milk. The principal remedy for the secondary effects is

HEPAR SULPHURIS,

(six globules,) dissolved in eight ounces of water, a table-spoonful of which must be taken every day.

NARCOTICS compose a large class of poisons: opium, and its preparations, nux vomica, tobacco, stramonium, cicuta, laurel, &c. Vomiting should be induced by four grains of tartar emetic dissolved in a tumbler of water, and after free vomiting, strong coffee should be copiously drank. In case of poisoning by opium, vinegar will be equally as beneficial as coffee.

NUX VOMICA or MERCURIUS

may be required for the consequences.

POISONOUS FISH, SHELL FISH, &c. The best antidote

is charcoal mixed with sugar and water, and afterwards smelling of camphor, or drinking strong coffee. If, after poisoning by the above, there should be an eruption, with swelling of the face,

BELLADONNA

may be given.

SULPHATE OF COPPER, IRON; AND ZINC. Tepid water, sweetened with sugar, or the white of eggs dissolved in water till vomiting is produced, and afterwards mucilaginous drinks.

CHAPTER XI.

HYDROPATHY, OR THE WATER CURE.

As hydropathy is, at the present time, engaging much attention, and as it is, to a certain extent, regarded by some practitioners as a valuable adjunct to homœopathy in expediting the cure of several diseases, a few particulars here presented respecting its origin, progress, mode of application, &c., may not be inappropriate. Long previous to the time of Priessnitz, the late distinguished hydropathist of Europe, water had been used as a remedial agent in various complaints; but its sytematic application to all forms of disease originated in the institution at Graefenberg, under the direction of the "Silesian peasant." The success which the new method there met with attracted public attention, and led to the establishment of similar institutions throughout Europe and America That it is a universal remedy; or that it is, as a system, at all comparable to homœopathy, we are far from believing; but as an occasional assistant to the action of homœopathic doses, as a depurative stimulating tonic and simple antiphlogistic, its importance is now widely acknowledged.

The modes in which this economical and easily attainable agent is applied are various, consisting of the wet sheet, the seat bath, foot bath, head bath, douche, bandages, &c. A brief description of these different modes of application are given below, although the principal, if not the only forms ever resorted to by the homœopathic practitioner, as an auxiliary merely to the internal use of medicine, are the wet sheet, the seat bath, and compresses.

THE WET SHEET.

This expedient is resorted to in febrile affections, when the surface is hot and dry. It assists in some instances the action of *Aconite*, and relieves the excessive restlessness accompanying fever, producing quiet sleep, and promoting perspiration. In inflammatory, eruptive, and other fevers, its judicious application has frequently been followed by the most favorable results, working in conjunction with the well-selected internal remedy to aid the recuperative efforts of Nature. The following is the process recommended: —

A sheet is to be dipped in cold water, and wrung out

as dry as possible; then spread upon the top of two blankets previously laid upon the bed or mattress. The patient, entirely undressed, is to be immediately laid upon the sheet, and closely covered with it, from the neck to the feet. After the entire body, with the exception of the head, is enveloped in the sheet, the blankets underneath are to be separately drawn over the body, and closely tucked in, one over the other. Particular attention should be paid in enclosing the neck, both by the sheet and blankets, that the cold air from without may not find an entrance. The lower ends of the sheet and blankets should be bound round and then drawn under the feet, so that, when the packing is completed, the whole surface of the body may be closely and entirely enveloped. Afterwards five or six thick blankets should be spread over the patient, and tucked under, or pressed closely against the sides. After the operation is thus concluded, one or two tumblers of cold water, drank slowly, will aid in bringing on perspiration.

At the end of one hour, — longer or shorter, however, as the perspiration may be more or less decided, — the patient should be carefully and quickly uncovered, and the whole surface rubbed with a towel or sponge dipped in cold water, and afterwards with a dry cloth.

This sponging with cold water while the body is in a state of perspiration, induced in this manner, is not, as many are apt to suppose, attended with danger. When perspiration is the result of violent, fatiguing exercise.

emotion, or certain feverish excitement, the sudden application of cold might be injurious. The person, packed as above, should not be forced to breathe the confined air of a closed room. A window in the apartment should remain open while the body is enveloped in the sheet.

Should the head become hot and uncomfortable, a towel, wet with cold water, may be laid upon the forehead. If, however, during the perspiration, the head should continue heated, or if there is general uneasiness, the blankets and sheet may be removed; such symptoms generally indicating a too powerful reaction.

When the wet sheet is applied to persons suffering from violent febrile action, hot, dry skin, and frequent pulse, the object is to abstract the superabundance of heat as rapidly as possible; and the application of a second wet sheet is necessary as soon as the first becomes dry, or the heat of the surface returns. As partial inflammation is subdued by the topical application of cold embrocations, so in this case, when the inflammation is general, the principle remains the same.

A half sheet is applied as a whole sheet above described, but extending only from the armpits to the thighs, and is both useful and convenient.

THE SEAT BATH.

The seat bath is taken in a tin vessel constructed for the purpose, with a back to lean upon; or a common tub may be made use of in an emergency. Such a quantity of water is to be poured into the tub as to reach above the hips of the person sitting. The upper part of the body, as well as the legs, should be covered. The length of time for remaining in this bath must depend upon circumstances; usually, however, from fifteen to twenty minutes, sometimes an hour. When the purpose is to counteract chronic congestion of blood, the time is frequently prolonged for two hours.

It is proper to add here that too protracted an application of cold water, either generally or locally, is unsafe; and the duration of time should not be left to conjecture, or to the judgment of the bather. In acute complaints, as brain, lung, nervous fever, or when violent pains exist, the time should be graduated according to the severity of the inflammation or pain.

The seat bath is used with much success, in connection with suitable internal remedies, for the relief of the pain, and the checking of discharges in diarrhœa, dysentery, bleeding, hemorrhoids, &c. In many cases, the relief is immediate and decided.

THE FOOT BATH.

Any vessel may be used for this purpose that will contain water sufficient to cover the feet as high as the ankles; and the bathing should not be continued more than fifteen minutes, the feet being kept in motion while in the water. As a derivative in affections of the head or chest, and as a remedy for an habitual cold

23*

condition of the feet, to which many persons are subject, this local application is frequently of great service.

THE HEAD BATH.

In this form of bath, the back part of the head is made to rest, for from ten to twenty minutes, in a narrow basin, containing about three inches of water; the body of the patient being extended on a mattress, with the shoulders elevated and supported by pillows. Or the water may be applied in another manner, by pouring from a pitcher upon the temples or back of the head, a vessel being placed underneath to protect the floor, and the shoulders covered by a blanket. In affections of the head, attended with increased temperature of the surface, the application of folded cloths, wet with cold water, will usually allay heat and pain; while the more powerful sedative action, from the processes above described, is thought to be required in cases of convulsions, apoplexy, &c. Cold water should never be allowed to fall on the head in the form of a shower bath when there exists much nervous susceptibility, or tendency to disease of the heart or lungs. In truth, the shower bath is but seldom employed, even in exclusive hydropathic practice, without long preparatory treatment.

THE DOUCHE.

A stream of water, forcibly projected from a hose or pipe from one to two inches in diameter, constitutes the

" Douche." It is used principally to reduce tumefaction by stimulating to action the absorbent vessels of
any particular part, as in chronic enlargement of joints,
certain kind of tumors, &c It should never be applied
to the head, nor for a long time to any portion of the
spine. Warm water has been used in this form with
advantage where the object has been to obviate muscular contraction, and in affections of a neuralgic nature,
accompanied by general nervous irritability, which will
not endure the application of cold.

COLD BANDAGE.

For the removal of local pain or inflammation, a
towel is to be wrung out dry in cold water, and applied
around or upon the part affected, and closely covered
with a perfectly dry cloth. Bound around the abdo
men in this manner, it produces perspiration, and
quiets the nervous irritability accompanying a feverish
condition of body. This is a beneficial application in
cramps and pains in the stomach and abdomen ; and its
frequent renewal is of great service in a constipated
state of the bowels.

In headaches, with external heat, a small, folded
towel, wrung out in cold water, and laid upon the forehead, will be productive of great relief.

In croup, the throat may be kept constantly bound
with a cold compress, closely covered with a dry cloth.
The local inflammatory action, which, unchecked, renders this complaint so dangerous, is more readily and

safely subdued by this simple method than by any other
external application.

The " Hydropathic Encyclopedia," by Dr. Trall, the
most extensive and useful of all works published on the
subject, contains observations on bandages, and general
directions for bathing, well worth insertion, and to
them the reader's attention is now invited.

" BANDAGES. These may be local warming or cooling processes, as indicated, and answer all the purpose
of the awkward, bungling, and expensive machinery of
liniments, lotions, poultices, embrocations, blisters, ru-
befacients, epispastics, cuppings, issues, burnings, and
other external drug appliances of the old school.

" A warming bandage, or compress, is simply one or
more folds of linen cloth, wet in cold water, applied to
the part affected, and covered with a dry cloth or other
material, to retain the animal heat.

" A cooling bandage, or compress, is a similar wet
application without the dry covering, or with the covering so light as to allow the animal heat readily to pass
off. In both cases the cloth is to be renewed as often
as it becomes dry. As usually managed, these compresses are both cooling and warming; the first impression being cold, and the reaction leaving a glow upon
the surface ; but they can be made to produce a constantly cooling effect, by very lightly covering and
frequently changing them ; or a very heating effect, by
covering them with flannel or other non-conducting
material.

" Coarse linen cloth, as common crash towelling, is

the most suitable cloth to be wetted; and for the dry covering, the same material, or any common muslin, will answer in warm weather, and soft flannel in cold weather. India rubber, gutta percha, and oiled silk have all been in repute, and a few years ago were very generally employed for coverings. I regard them all as objectionable. They do, indeed, serve to prevent evaporation, and retain more perfectly the animal heat; and they also keep the part moist longer; and they seem, too, to have a more drawing or derivative influence, if the more ready production of eruptions or boils indicates such influence. But they retain the effete, perspirable matter which should pass off; and their non-conducting or non-electric property renders them relaxing and weakening to the cutaneous function.

" It seems to me that, *in all cases*, cloth coverings are the best. If they produce a less number of boils, or less painful eruptions, the cure will nevertheless be as prompt, and even more perfect. When the skin is torpid and cold, canton, or soft, light, woollen flannel, answers every purpose; and, if necessary, for very feeble patients, who are unable to take much exercise, two or three thicknesses may be used.

" THE CHEST WRAPPER. This is advantageously employed in nearly all chronic diseases of the chest, as, incipient consumption, bronchitis, in the very early stage of hydrothorax, or dropsy of the chest, spasmodic or periodical asthma, &c. It may be made of crash towelling, or two or three folds of muslin, and fitted with armholes loosely to the trunk of the body, from the

neck nearly or quite down to the hips. The outside
covering is a similar wrapper, made of the same material or of flannel. The inner or wet wrapper is tied as
tightly around the body as desired, by tapes, which are
attached to the top, bottom, and middle; and the outside or dry wrapper is either tied around it, or the inner
one is buttoned to the outer.

" There is some discrepancy in the views of different
hydropaths, as to whether the wet cloth should extend
entirely around the body, or a few inches over the spine
be left uncovered. Here, again, as in most of the
vexed questions which occur in hydropathic bathing,
the feelings of the patient are our best guide. If the
wet cloth over the spine does not produce any disagreeable chilliness, pain, or uneasiness different from what
is experienced when the partial wrapper is worn, I
would have it entirely encircle the trunk; otherwise a
space of from four to six inches in the centre of the
back should be left covered by the wet cloth.

" This may be worn day and night for several weeks,
provided it produces no uncomfortable chilliness during
the day, and does not become so warm and dry as to
make the patient restless during the night. In the
former case, it should only be worn during the warmest
part of the day, or during time allotted to exercise, or
from the forenoon bath until evening. In the latter
case, it may be worn during the day, and omitted at
night. It usually requires wetting, when worn constantly, in the morning, towards noon, towards evening,
and at bed time.

"THE ABDOMINAL WRAPPER. — The wet girdle, or abdominal compress, as this is generally called, is more frequently employed than any other local hydropathic application. Derangements of the digestive organs are so prevalent now, that those who do not thus complain are exceptions to the general rule; and for all these complaints this bandage is appropriate. It is also serviceable in all chronic diseases of the liver; and in acute diseases of the abdominal viscera, as inflammation of the stomach and bowels, cholera, dysentery, cholera morbus, diarrhœa, it is always employed with benefit.

" A great deal of ingenuity has been wasted in con triving abdominal compresses. But the best invention of all is three yards of common crash towel cloth. One half of this is wet, and moderately wrung; the wet end is applied to the side of the abdomen; then the bandage is passed across the abdomen, and around the body, followed by the dry half. This brings two folds of the wet part over the front of the abdomen, and one behind. Whether it is to be extended entirely around the body must be determined by the rule mentioned as applicable to the chest wrapper. The proper crash cloth is from twelve to sixteen inches wide, and covers the trunk from the short ribs to the hips, descending a little over the latter. As with the chest wrapper, it may be worn constantly c occasionally. It should never be applied so tightly as to hinder, in the least, free perspiration. It may be kept in place by tapes or pins.

" GENERAL RULES FOR BATHING. — 1. No bath should
be taken on a full stomach. General baths, as the wet
sheet, plunge, douche, shower, &c., should not be taken
until the process of digestion is nearly or quite completed — from three to four hours after a full meal.
Local baths, as the foot, hand, &c., may be taken in an
hour after a light, and two hours after a hearty meal.
Bandages may be applied at any time.

" 2. Patients should not eat immediately after a bath.
An hour is soon enough after a full, and half an hour
after a local bath.

" 3. All patients who are able should exercise moderately previous to a bath, unless at the bathing time the
body is in a warm glow ; and after a bath, according to
muscular strength. The more exercise, short of absolute fatigue, the better. By absolute fatigue, I mean
that degree of exhaustion which is not readily recov-
ered from on resting.

" 4. In very warm weather, the most active exercise
should be taken before breakfast ; and, during the heat
of the day, it should not be beyond what is perfectly
agreeable.

" 5. No strong shock should ever be made upon the
head. A shower or pail douche, poured, but not dashed
on, is not objectionable for those who enjoy a tolerably
well-balanced circulation, and are not subject to nervous
headache.

" 6. Profuse perspiration, or great heat of the body,
is no objection to any form of cold bath, provided the
body is not in a state of exhaustion from over-exertion,.

nor the breathing disturbed. This point is generally misunderstood by physicians, and medical books of the old school are wholly in error about it. The majority of people imagine that the sudden transition from cold to hot is dangerous. The danger is all on the other side — applying cold when the body is already too cold. Again, it is thought that a cold bath, when the body is dripping with sweat, will check the perspiration, and do immense mischief by driving it in. This is a mere fantasy. The matter of perspiration is a viscid, waste, dead, effete material, and its presence on the surface has nothing whatever to do with the effect of a cold bath. It may be as safely washed off with cold water when the body is hot, as can any other extraneous matter adherent to the surface.

" But persons are often injured by going into cold water when the body is hot and perspirable. Granted. I have known several young men made cripples for life by this practice. Now, what is the explanation? Either the body was too cold, or in a state of exhaustion, or the respiration was materially disturbed, or the stomach was loaded, or all of these conditions existed together. There is a reciprocal relation between circulation and respiration, which cannot be greatly disturbed without injury. If a person jumps into cold water when out of breath from violent exercise, he endangers his health, because the intimate sympathy between the action of the heart and lungs will prevent reaction to the surface ; and the result is internal congestion. Under all other circumstances, a warm or hot skin is favorable to any

24

cold application, while the state of perspiration is a matter of no sort of consequence, one way or the other. Dr. Johnson remarks, ' Being in a state of perspiration is no objection to taking any bath, except the sitz, foot, and head bath.' If the rules I have laid down are duly observed, there can be no force in the objection of Dr. Johnson.

" 7. When full treatment is prescribed, as three, four, or five baths a day, the patient should take the most powerful, or those which produce the greatest shock on rising, and in the early part of the day.

" 8. Wetting the head, and even the chest, is a useful precaution before taking any full bath, and especially important for patients who are liable to head affections."

DRINKING OF COLD WATER.

The following remarks respecting the frequent and excessive use of cold water as a beverage, so highly recommended by some practitioners of hydropathy, are from the work of Dr. Edward Johnson, the most eminent hydropathist in England. He writes thus: " I am decidedly opposed to the indiscriminate drinking of large quantities of cold water. One cannot understand in what manner these large imbibitions are to operate so as to be useful in the animal economy. We know precisely what becomes of the water, soon after entering the stomach ; we can trace exactly what course all this water must take, what channels it must traverse, between its entrance and its exit. We are perfectly well acquainted* with certain physiological effects produced by it, after it has been received into the system.

"It dilutes the blood; it lowers the temperature, and thereby diminishes the vital power of the stomach; it puts certain systems of capillary blood vessels on the stretch, to the great danger of bursting; and it overtaxes the kidneys. I have seen two cases of bloody urine, which were fairly attributable to the excessive drinking of water.

"An unfortunate gentleman of Nottingham, England, who died from excess of treatment, administered by himself, was found to have the fine, thin, transparent mucous membrane of the stomach semi-dissolved into a gelatinous pulp (which was easily scraped off) by the quantities of water he had drank. He had been covered with boils, had a most ravenous appetite, and had drank seven or eight pints of water daily.

"It must be remembered, that, in drinking cold water, the full shock of the cold is sustained by the stomach alone. It is from that organ that nearly all the heat is abstracted by the cold water. While the water remains in the stomach, it is continually abstracting vital heat from it. The water warms itself by heat abstracted from the stomach. When it leaves that organ and enters the system, it has become warm water; and the heat, which it has absorbed from the stomach into itself, it carries away into the blood vessels, leaving the stomach chilled, and with a lower temperature than any other part of the body. This lowering of its temperature, repeated frequently, has a decidedly weakening effect upon the stomach. The capillary blood vessels, deprived of their vital heat, become relaxed; they open, and admit a larger current of blood; congestion thus takes place; irritation is set up, like that in a bloodshotten eye; and a morbid craving for food, even between meals, is produced.

" If the water imbibed, indeed, lowered the temperature of the whole body equally, the case would be different, and the practice less hurtful.

" Thus, then, it seems there are certain well-understood and very obvious injuries, which the large imbibition of water cannot fail to inflict; while the supposed benefits to accrue from it are altogether mystical, problematical, unintelligible. This, however, only applies to excessive drinking — drinking for mere drinking's sake, as one formerly swallowed physic. If persons are thirsty, if their mouths and stomachs are heated and feverish, let them drink as much water as is sufficient to allay these uneasy feelings. If the tongue be foul in the morning, and the mouth parched, half a tumbler of pure, spring water will be found very refreshing, and provocative of an appetite for breakfast. The quantity of water which each person should drink during the day must always depend on his own feelings He may always drink when the doing so is agreeable to his sensations; when it is repulsive, *never.*

"A large quantity of fluid should not be taken during dinner. It should not exceed a tumbler full; and the less the better, provided a proper quantity of food can be got down without it. A natural thirst will occur some three or four hours after dinner, and then a hearty draught of cold water will be delicious and useful.

."All the intelligible good effects of water drinking will be as certainly obtained from drinking some six or seven tumblers a day (including meals) as by drinking more; while all the evils of *excessive* drinking will be avoided.

"Whenever the appetite is deficient, I recommend the patient to drink a tumbler or two of fresh, cold water before breakfast, and two before dinner; and to take cold water for breakfast and supper instead of tea, if it do not *disagree* with the stomach."

There are various kinds of water, all possessing different degrees of purity; and, as it is of importance to be made acquainted with those best adapted for drinking, we again quote, from the excellent work of Dr. Trall, already alluded to, the following remarks on this subject:—

DIFFERENT KINDS OF NATURAL WATERS.

"The natural waters of the globe have been classed into *common waters,* comprising rain, spring, river,

24 *

well or pump, lake, and marsh waters; sea waters, including the ocean, and the salt lakes, or inland seas; and mineral waters, to which class belong all the springs, streams, or pools usually regarded as medicinal.

"Rain water is the purest of all natural waters. When collected in cities, it is more or less impure at the commencement of the shower, from admixture with foreign matters suspended in the atmosphere, and is often loaded with particles washed from the roofs of the buildings. After several hours of continuous rain in cities, and a much shorter time in country places, it comes down almost perfectly pure. Air is a constant constituent of, or admixture with, rain water; and it contains a slight trace of carbonate of ammonia, which is probably a product of animal decomposition, and the cause of rain water so readily running into the putrefactive process. *Snow water* does not differ materially from rain water, except in not containing air. That it is injurious to health has long been a vulgar error. Eating snow, however, does not quench thirst; but melted snow is as efficacious for this purpose as rain water.

"Spring water only differs from rain water in having percolated through the earth, and having, during its passage, either imparted·some of the particles it held in solution to the soil, or taken up soluble matters from the soil, or both. Its properties will therefore depend entirely upon the nature of the soil. A majority of the springs in the United States are *hard*, owing to earthy and saline matters, the most common of which are sulphate and carbonate of lime. There are, however, many *soft water* springs ; enough, in fact, to answer all the drinking purposes of as dense a population as the country can sustain, if it were conveyed to, and distributed among, the dwellings. The people in the country are generally singularly inattentive to the important matter of providing themselves with *pure soft water.* They are very apt to get their supply from the most convenient spring, instead of the best. If they fully appreciated the importance of good water, they would not locate the dwelling house until they had located the spring or well.

" River water is an admixture of rain and spring water ; it always holds in suspension a greater or less amount of extraneous matter, and, in and around cities, is strongly contaminated with decomposing animal and vegetable matters. Much of the river water in this country, as it runs through sparsely populated districts, is comparatively quite pure and healthful.

" The water of the Thames, and in the vicinity of London, contains, as impurities, about twenty grains of solid matter to the gallon. Of this, carbonate of lime constitutes about sixteen grains, and sulphate' of lime and common salt about three and a half grains.

" The Croton water of New York contains but a trifle over four grains of solid matter to the gallon, only a grain and a half of this being carbonate of lime; sulphate of lime, the chlorides of calcium and magnesium, and the carbonate of magnesia, constitute a little over two grains. The Cochituate water of Boston is equally as pure; and the Schuylkill of Philadelphia nearly as pure.

"Previous to the introduction of the Croton River, the Manhattan water supplied to the citizens (of New York) contained, in Chambers and Reade Streets, 125 grains of impurities to each gallon; in Bleecker Street, 20 grains; and in Thirteenth Street, 14 grains. Some of the wells in the lower part of the city contained 58 grains. The water in the wells of Boston and Philadelphia were in no better condition.

"The usual results of drinking very hard waters, and those strongly impregnated with the exuviæ of animal and vegetable substances, are severe dysenteries or protracted diarrhœas, and chronic affections of the kidneys.

"Well water is generally more impregnated with earthy salts, especially bicarbonate and sulphate of lime, than river water, or even spring water. Its hardness is shown by its curdling and decomposing soap, instead of mixing with it readily, and forming a suds, as will soft water. Sulphate of lime (gypsum, plaster of Paris) is a frequent cause of diarrhœa.

"Horses manifest such an instinctive repugnance to hard water, that they will drink out of a turbid or muddy pool, provided its water is soft, in preference to partaking of the clearest and most transparent water, if it is *hard*.

" Lake water is generally very impure, being a collection of rain, river, and spring water, contaminated with putrefying animal and vegetable matters.

" Marsh water is similar to lake water, but still more loaded with offensive and putrescent organic matters. The stench arising from marshy and swampy grounds, which are occasionally inundated from the sea, is owing to the decomposition of the sulphates of the sea water, by the putrefying vegetable matters; which process evolves the intolerable sulphuretted hydrogen gas.

" Sea water contains, on the average, three and a half per cent. solid matter. The amount varies considerably in different seas, and in different parts of the same sea. Its composition also varies in different localities. An analysis of 1000 grains of the water of the Mediterranean gave the following result: Water, 959.26; chloride of sodium, (common salt,) 27.22; chloride of potassium, 0.1; chloride of magnesium, 6.14; sulphate of magnesia, 7.02; sulphate of lime, 0.15; carbonate of lime, 0.20. Iodine and bromide of magnesium have been found in some sea waters.

" Taken into the stomach, sea water excites thirst, nausea, and, in large doses, vomiting and purging.

" Mineral waters are classed according to the character of their prevailing impurities. Those whose predominating active principle is iron, are called *chalybeate*, or *ferruginous*. *Sulphurous* or *hepatic* waters

are strongly impregnated with the offensive sulphuretted hydrogen gas. *Carbonated* or *acidulous* waters
contain carbonic acid, which renders them sparkling
and pungent. Of the *saline* mineral waters, there are
many sub-varieties, as the *calcareous, alkaline, sili-
cious,* &c.

"The *medicinal* fame of the 'Congress water,' at
Saratoga, is derived from the great amount of its
deleterious ingredients. One gallon contains the following impurities: Chloride of sodium, 3850. grains;
hydriodate of soda, 3.5 do. ; bicarbonate of soda, 8.982
do. ; bicarbonate of magnesia, 95.778 do. ; carbonate
of iron, 5.075 do. ; silex, 1.5 do.; hydrobromate of
potash, a trace; in all, 597.943 grains. Each gallon
also contains 311 cubic inches of carbonic acid gas.
and seven of atmospheric air.

"Dr. Steel, of Saratoga, very judiciously advises
those who wish to experience the full benefit of this
water, to drink it *only once a day* — about three pints
early in the morning; and he remarks, very sensibly,
'It would be much better for those whose complaints
render them fit subjects for its administration, if the
fountain should be locked up, and no one suffered to
approach it after the hours of nine and ten in the
morning. If it should be locked up at all hours of
the day and night, and a stream of pure, soft water
substituted, the advantage to the invalid portion of the
guests would be still greater.'

"The Iodine Spring, at that place, differs from the
former mainly in containing three and a half grains

of iodine to the gallon, with a little more than half the quantity of the other ingredients. The Sans Souci Spring, at Ballston Spa, differs from the Congress principally in containing carbonate of lime instead of bicarbonate of magnesia, and possessing altogether a little less than half the amount of impurities."

TESTS OF ORDINARY IMPURITIES.

The following are the tests (from Pereira's "Food and Diet") by which the presence of the usual impurities of common water may be ascertained : —

" 1. *Ebullition.* By boiling, air and carbonic acid gas are expelled, while *carbonate of lime*, held in solution by the carbonic acid, is deposited. This deposit is the fur or crust which lines tea kettles and boilers.

" 2. *Protosulphate of Iron.* If a crystal of this salt be introduced into a vial filled with the water to be examined, and the vial be well corked, a yellowish-brown precipitate (sesquioxide of iron) will be deposited in a few days, if *oxygen gas* be contained in the water.

" 3. *Litmus.* Infusion of litmus, or syrup of violet, is reddened by a *free acid.*

" 4. *Lime Water.* This is a test for *carbonic acid*, with which it causes a white precipitate, (carbonate of lime,) if employed before the water is boiled.

" 5. *Chloride of Barium.* A solution of this salt usually yields, with hard water, a white precipitate, insoluble in nitric acid. This indicates the presence

of *sulphuric acid*, which, in common water, is com
bined with lime.

"6. *Oxalate of Ammonia.* If this salt yield a white
precipitate, it indicates the presence of *lime*, carbonate
and sulphate.

"7. *Nitrate of Silver.* If this occasion a precipitate
insoluble in nitric acid, the presence of *chlorine* is
inferred.

"8. *Phosphate of Soda.* If the lime contained in
common water be removed by ebullition and oxalic
acid, and to the strained and transparent water ammonia and phosphate of soda be added, any *magnesia*
present will, in the course of a few hours, be precipitated in the form of the white ammoniacal phosphate
of magnesia.

"9. *Tincture of Galls.* This is used as a test for
iron, with solutions of which it forms an inky liquor,
(tannate and gallate of iron.) If the test produce this
effect on the water before, but not after boiling, the
iron is in the state of *carbonate;* if after, as well as
before, in that of *sulphate.* *Tea* may be substituted
for galls, to which its effects and indications are similar. *Ferrocyanide of potassium* yields, with solutions
of the *sesqui-salts of iron*, a blue precipitate, and with
the *proto-salts* a white precipitate, which becomes blue
by exposure to air.

"10. *Hydrosulphuric Acid*, (sulphuretted hydrogen.) This yields a dark (brown or black) precipitate
(a metallic sulphuret) with water containing *iron* or
lead in solution.

" 11. *Evaporation and Ignition.* If the water be evaporated to dryness, and ignited in a glass tube, the presence of *organic matter* may be inferred by the odor and smoke evolved, as well as by the charring. Another mode of detecting organic matter is by adding *nitrate* or *acetate of lead* to the suspected water, and collecting and igniting the precipitate, when globules of melted lead are obtained, if organic matter be present. The *putrefaction* of water is another proof of the presence of organic matter. *Nitrate of silver* is also a test, as before mentioned.

"PURIFICATION OF COMMON WATER.

" *Filtration* removes all insects, living beings, and all suspended impurities; but it does not deprive water of the substances it holds in solution. *Boiling* destroys the vitality of any animals or vegetables it may contain, expels air or carbonic acid, and causes the precipitation of carbonate of lime. Sometimes it may be advantageous to *boil* water first, and *filter* it afterwards. Distillation purifies water from every thing except traces of organic matter; it is, however, a process too troublesome and expensive for general employment. *Chemical agents* are sometimes made use of to free water from particular ingredients. Alum, two or three grains to a quart, will cleanse muddy water: the alum decomposes the carbonate of lime; sulphate of lime is found in solution; and the alumina is precipitated in flocks, carrying with it mechanical

25

impurities. Though this process renders the water
clear, it adds nothing to its healthfulness, but renders
it even harder, by converting the carbonate into sulphate of lime. *Alkaline carbonates* soften water by
decomposing all the earthy salts, and precipitating
earthy matters. The carbonates of soda and potash
are much used in washing, on this account; they do
not render the water any purer — not fit for drinking
or culinary purposes.

"ADULTERATIONS OF COMMON WATER.

" The purest water is liable to become impregnated
with poisonous properties, when conveyed through
some kinds of metallic pipes, particularly leaden ones.
The air contained in very pure water rapidly corrodes
lead. Distilled water, from which the air is excluded,
has no action on it until air is again admitted, when a
thin, white crust of carbonate and hydrate of the oxide
of lead is speedily formed. Rain water is often impregnated from the lead of roofs, gutters, cisterns, and
pipes. Combinations of lead, iron, and zinc, and other
mixed metals, — as in cases where iron bars are used
to support leaden cisterns, the introduction of iron
pumps into leaden cisterns, &c., — often produce a
galvanic action which dissolves a portion of the lead.
The leaden covers of leaden cisterns are also a source
of contamination; the water evaporates from the cistern in the form of pure or distilled water, and con-
denses upon the lid, which it corrodes, and then falls

back into the cistern, impregnated with the metal. Such cisterns should have wooden covers.

" Various saline matters impair the corrosive action of water and air, and exercise a protecting influence. The carbonates and sulphates afford the best security against lead poisoning, because they form a protecting crust upon the surface of the metal. Dr. Lee declares that ' palsy is often met with in the city of New York among grocers and porter house keepers, and is doubtless owing to their drinking beer in the morning which has stood in the lead pipes over night.'

" Chemists do not agree respecting the action of our Croton water on its leaden conduits; but experience settles the question affirmatively. It becomes our citizens, therefore, to exercise a constant watchfulness in its employment, which is, to let as much water run as the leaden pipes contain to their junction with the iron pipes in the streets, before drinking it. With this precaution and the frequent emptying of the leaden pipes through the day, it is not probable that any appreciable injury will be experienced from the lead. But these facts prove that the principle of conveying water through our dwellings by leaden pipes is wrong, and a substitute should engage the attention of ingenious men and philanthropists." — *Hydropathic Encyclopedia.*

CHAPTER XII.

REMARKS ON HYGIENE.

1. On Food.
2. On Clothing.
3. On Bathing.

4. On Air.
5. On Exercise.

THE subject of hygiene, or the art of preserving health, although not in intimate and necessary connection with therapeutics, or the treatment of disease, is of such manifest importance that no apology need be advanced on account of its introduction into a treatise intended for domestic circulation. The few remarks offered on this unquestionably useful, yet greatly neglected subject, will be comprised in five divisions, under the general heads of *Food*, *Clothing*, *Bathing*, *Air*, and *Exercise*. And first, of

ON FOOD.

In the chapter on Diet are enumerated the articles of food, which, by reason of their medicinal properties, interfere with the operation of homœopathic remedies; and a list of such articles as are allowed during medical treatment is also added; and, inasmuch as the prescribed regimen is in consonance with actual experiments establishing facts in relation to the comparative digestibility of food, it is well adapted for general consultation.

Judging from the structure of the human teeth, and the internal organization, the Creator evidently intended that man should derive sustenance from animal as well as vegetable substance; and wherever the race is most favorably situated for the convenient acquirement of both, as in temperate regions, more human health is found to be in the best condition, and national power and prosperity the most remarkable. In order that aliment, of whatever nature it is, should be properly prepared for the purpose of nutrition, it must first be subjected to the process of mastication or chewing, and mixed with the saliva in the mouth; and, after being swallowed and having reached the stomach, it is exposed to the action of the gastric juice, there freely secreted; and, by the solvent power of this peculiar secretion, it is converted into a soft, pulpy mass, which has received the name of chyme.

In this state, it is urged by muscular contraction into the duodenum, or passage leading from the stomach. Being there further mixed with bile and other fluids, it is converted into chyle. This chyle is the nutritious portion of the food swallowed. Numerous very delicate vessels now take up the fluid, thus eliminated from the food, and carry it

25 *

through glands, where it receives additional nutritive ·
properties.   Thence it is propelled through canals,
until it becomes mixed with the blood; and thus is
completed the process of digestion.   The first operation upon the food is, as has been mentioned, mastication; and this should be performed *deliberately*,
for, unless that which is swallowed has been well
chewed and mixed with the saliva, all the particles
of the food will not be exposed to the action of the
gastric fluid.   Digestion in its first stages being impeded, the whole process will be imperfectly executed,
and injury to health be the consequence.   The second
action upon the food is in the stomach; and that this
should proceed uninterruptedly, *mental and physical
repose* is needed; for much exercise of the mind and
body withdraws from the organs of digestion that
nervous energy and flow of blood which are especially
required by them while in a state of activity.   One
nour, at least, after a meal should elapse before active
muscular or mental exertion be permitted.

  Healthy digestion of food, then, depends upon its
minuteness of division, effected by slow and thorough
mastication, and upon repose after a meal.   Further
requisites are a proper amount of muscular exercise,
when the contents of the stomach have been sufficiently
acted upon by the gastric juice, and a suitable quantity
as well as quality of food.   With regard to the latter,
the dietary system alluded to in this work, while composed especially with reference to the invalid, is nevertheless adapted for the healthy, and is sufficiently

ample and various in the articles enumerated to satisfy the requirements of nature. Artificial wants, vitiated tastes, are not to be consulted; for it is undeniable that the greater the departure from a simple diet, the more uncomfortable and diseased are both mind and body.

The quantity of aliment is quite as important a subject for consideration as the quality. Two pounds daily has been estimated as the average amount of solids requisite for health; but the habits, mode of life, and other circumstances, render a slight departure from this amount, at times, advisable. As a general rule, they, who lead sedentary lives need not deviate from the above-named quantity; but they whose occupation requires much physical labor may exceed this amount with impunity. Much more is usually consumed, and it is certain that uneasy sensations of oppression, unfitness for thought and action, ill temper, and the thousand evils of indigestion, are the legitimate, direct consequences of the consumption of more food than nature requires. Lewis Cornaro, an Italian, who, at the age of eighty-three, wrote a treatise in commendation of temperance, states, that when forty years old, in consequence of impaired health, he adopted a strict, abstemious regimen, consisting of twelve ounces daily of solid food, with an equal quantity of liquid; and, as the result of such moderation, he enjoyed a remarkable degree of health and activity. He died at the age of ninety-eight.

It is not easy to designate clearly the proportion of

animal to vegetable food which may be individually suitable, since much depends on climate, habits, age, temperature, &c. Yet it may be stated, in general terms, that more animal food is required in winter than in summer; more in cold than in warm climates; more by those who labor out of doors than by the sedentary; by the old than by the young. In temperate latitudes it should constitute the smallest proportion of the quantity of food taken, particularly by the inhabitants of a crowded city; for, without sufficient exercise in pure air, much animal food tends to the production of serious disorders.

The digestive organs are frequently disordered by immoderate and hasty eating. And when from such causes, as well as others, an unhealthy condition is present, no general dietary rule can be made applicable, and experience is the best teacher with regard to the particular kinds of aliment most suitable in the sufferer's own case; for articles of food which are found to be easy of digestion with some, produce, in others, sensations of oppression and distress. The study of the comparative digestibility of food is, however, of much more use and interest to those whose powers of digestion are impaired, than to the vigorous and healthy; for to the latter the kind of food is not of so much consequence as its variety and amount. Daily restriction to one description of food, whether animal or vegetable, is not so conducive to health as when some variation is observed; neither is nutriment in a condensed form so well calculated for proper digestive action as when a similar amount of nutriment in greater volume is taken.

The lavish use of condiments is a frightful source of derangement and debility of digestion. Highly spiced aliment increases digestive action; and this increased activity is followed, as is the law with all stimulants, by a decrease of natural energy; so that when the stomach is called into exercise during this period of exhaustion, as it must necessarily be, it is incapable of performing its office in a healthy manner; and indigestion, with its long train of evils, results from this foolish method of provoking appetite.

The *time* for eating is another important point for consideration. The interval between meals should not exceed six hours; and regularity in this respect is essential to the maintenance of health. The Germans usually sit down to four meals during the day; the French to two; while in this country three meals are ordinarily taken; and such a division of the waking hours appears best adapted for health. Before the breakfast hour, no active exercise of mind or body should be engaged in; as, on the supposition that supper has been taken three hours before retiring to rest, and seven hours have been devoted to sleep, the system is not in a proper condition to endure much expenditure of strength previous to the morning meal. It is a common but erroneous impression, that, after a night's deprivation of sustenance, violent pedestrian exercise, or close application to study, can be better borne than at other periods of the day.

As to liquid nutriment, it is hardly necessary to
remark that none can supply the place of water. The
demand for artificial beverages is an unnatural, vitiated
appetite. Excepting under certain conditions of physical debility, all liquids other than that which Nature
abundantly furnishes are useless for the purposes of
nutrition, and most of them are positively injurious.
The total abstinence from stimulants of every description, under all circumstances, we are not disposed to
advocate ; for cases may and do occur when artificial
stimulation is really required for the restoration of
exhausted energy ; and the excitement consequent on
the use of alcohol, by which dangerous debility has
been and is yet to be averted, cannot be produced by
any other known agent. In the decline of life, for
example, when exhaustion of strength results, not
unfrequently, from slight exertion or mental emotion,
temporary restoration is effected by the stimulus of
alcohol in some form ; and healthy reaction, also, from
the depressing consequences of miasmatic exhalations,
which water alone has but little power to induce, may
at times be promptly brought about by a resort to artificial stimulation. It is to be understood, however, that,
in the instances alluded to, their use is remedial, and
justifiable only as a temporary expedient ; for no truth
is more clearly established than that *continued* artificial
excitement lessens the capacity of vital resistance to
depressing agents of every kind and degree.

ON CLOTHING.

The covering which is required · as a protection against the vicissitudes of the weather in most regions, is frequently, through ignorance or carelessness, made the means of producing a great amount of suffering. It is easier to declaim against than to resist the tyranny of fashion; and health will doubtless continue to be sacrificed, constraint and uneasiness endured, to the end of time, notwithstanding all that has been or may be written in relation to it. No costume could be contrived more uncomfortable, unhealthy, and unbecoming, than that which is in use at the present day, from the tight compresses about the neck, to the narrow, unyielding covering for the feet. The circulation of the blood is impeded, the free action of the muscles hindered, respiration disturbed, all of the important viscera bruised and disordered, by external pressure. Stocks, padded neckcloths, close-drawn vests, tight waistbands, corsets, and rigid shoe leather, are used in bold defiance of the laws of health, while there is also exhibited thereby a remarkable indifference to comfort and convenience. As the absurdity as well as danger of thus cramping the body must be obvious to all, while the prevailing fashion will be generally followed at all hazards, it seems useless to continue the subject; and we will pass to the consideration of the material that is most suitable for clothing.

The important properties to be regarded in the mate-

rial selected for inner garments are, the capacity of con-
ducting heat, of imbibing perspiration, and of mechani-
cally acting upon the surface of the body.

Flannel best fulfils the above-named conditions.
Being a poor conductor of caloric, it retains the animal
heat thrown off from the surface longer than any other
substance; and its porous texture causes it to imbibe
the insensible perspiration that is constantly exhaled
from the skin ; while, by its action upon the cutaneous
vessels, a sufficient excitement is induced to counteract
the depressing influence of cold in our climate.

Next to flannel may be classed *silk*, which possesses
the required properties of porosity, and is a non-con-
ductor of caloric; but the smoothness of its surface
prevents the exciting action upon the skin alluded to as
a desirable quality. Its texture also in the common
fabrics is too open for security against exposure to
sudden changes of the weather ; and it is consequently
better adapted to a more equable and milder temperature than prevails in the changeable climate of New
England.

Linen and *cotton* are garments suitable for warm
countries, or for the summer season in temperate latitudes ; but both — more particularly the former — are
poorly adapted for protection against the severity of
winter. Hosiery of either material should not be worn
more than three months in the year ; as warm and dry
clothing for the feet is of the utmost importance.

There are various kinds of delicate leather in use for
inner garments, especially among the inhabitants of the

western portion of our country, as the skins of deer,
hare, &c., which serve as excellent preventives of cold,
but are not suitable for warm weather. The rapid
alternations from heat to cold in our climate require
the use of some substance which, under all conditions
of the weather, will answer the purpose both of comfort and security. Woollen, in its different forms of
manufacture, is preferable in every respect to any other
material.

The raiment used, whatever is its character, ought to
be of uniform thickness, and cover all portions of the
body equally; for nothing is more unsafe than partial
exposure. The chest and abdomen should be carefully
guarded, as the sensitive organs situated within these
cavities are peculiarly liable to inflammatory affection.
The diminution of warmth and of nervous action on the
surface results in a proportionate increase of action
internally. The capillary vessels through which the
blood flows, in its passage from the arteries to the veins,
are contracted by cold; and the blood, being repelled
from the circumference, and prevented by the continued
application of cold from freely circulating in the capillaries, centres in the cavities, which contain organs
essentially vital. The consequence is inflammation of
a very dangerous character.

Clothing is therefore to be employed as a protection
against cold; the whole person being covered with
flannel during those seasons, especially, when sudden
changes are apt to occur; and there are but two or
three months of the year, with us, during which this

26

article of dress can be safely dispensed with. It is by
no means prudent to go very thinly clad either by day
or by night; for sudden and great variations not unfrequently take place even in midsummer, exposing such
as are insufficiently protected to dangerous attacks of
inflammation, and to that destructive complaint of the
lungs against whose insidious approach too much precaution with regard to dress cannot be directed. An
uncomfortable degree of warmth promoted by clothing
is, on the other hand, to be avoided; for extreme susceptibility to atmospheric influences may thus be induced, predisposing the system to the inroads of disease
on the slightest exposure. Suitable attention to one's
own sensations, preserving an agreeable medium
between the two extremes of undue artificial warmth
and a painful sense of cold, is the safest course to be
pursued.

Particular attention ought to be given to the clothing of children. During infancy, the power of genera-
ting heat is less than at any other period of existence,
with the exception of extreme old age. In early life,
the circulation is mostly confined to the surface, the
internal organs being in a comparatively feeble condition; and the great mortality among infants is in part,
doubtless, justly attributable to a deficiency of clothing.
In consequence of the rapid and unforeseen changes to
which this climate is liable, thin dresses cannot, with
much safety, be adopted and discontinued at given
periods; and neither is it always proper that the same
dress be worn during both morning and evening. Very

young children, however warm may be their clothing, cannot breathe with impunity extremely cold air, as their lungs are not yet fitted for a low temperature, and exercise is needed to promote equal circulation. They should, therefore, not be carried out of the house during the severity of winter.

ON BATHING.

In order that the importance of bathing may be fully understood, it will be necessary to allude particularly to the function of the skin. A watery secretion exudes through every portion of the exterior membrane of the body, from the glands situated under and near the surface; and this exudation is going on perpetually, in an imperceptible form usually, but made visible when the circulation is accelerated by any cause. Nearly two pounds, either as fluid or vapor, of waste matter, which is thus thrown out of the system, passes through the skin daily. This waste matter is composed of water and various salts; the former being evaporated, while the latter ingredients are retained on the skin, there to accumulate and impede the further egress of perspiration, unless removed by ablution. To promote free perspiration, therefore, by preventing the collection of saline particles upon the surface, which interferes with a process indispensable to a healthy condition of the body, must obviously be of the greatest importance.

While in the act of breathing, vapor is exhaled from

the lungs; and, if the action of the skin be interrupted, this exhalation must be increased. In such a case, respiration becomes laborious; and over-exertion of the lungs, thus caused, tends to debility of that organ, and induces consumption, with other diseases of the chest. Consequently, to offer every facility for the promotion of unobstructed perspiration, is to adopt a precautionary measure against the attacks of pulmonary disease, which, in our climate, needs but little encouragement. The escape of the evaporable portion of the perspiration may be facilitated, as was stated in the preceding article, " On Clothing," by porous substances worn next the skin, and prevented,・on the contrary, by material that is impermeable to air and moisture. But the impurities which do not evaporate, and which constantly tend to accumulate, must be removed by water; and that removal should be effected once, at least, every day.

Water acts favorably, not only by cleansing the skin, but, by its secondary effect, increasing the action of the cutaneous vessels, strengthening the nervous system, furthering vigorous circulation, and rendering the body more capable of enduring, with impunity, the rapid atmospheric changes to which it may be exposed. The immediate effect of its application is depressing. The action of the capillaries, and that of the nervous and vascular systems, are diminished. When reaction takes place, which is indicated by the sensation of warmth, all the animal functions are carried on with renewed energy, and its happy influence is manifested by the reanimation and increased capability of both physical and mental powers. Should this reaction occur early, and in a marked manner, the application of cold will have proved beneficial. If, on the contrary, the primary action continues long, the sensation of chilliness not being followed by that of warmth, but accompanied by debility and exhaustion, the bathing is of doubtful advantage, or rather of decided injury. It is evident that, where reaction does not take place, — where there is a deficiency of vital activity, from whatever cause arising, — the influence which adds to the pre-existing depression of the powers of life, and is prolonged in its action, must be an unfavorable one. Reaction does not readily occur in the debilitated and delicate, nor in infancy and old age, when the vital functions are feebly executed. It may be favored, however, when thus delayed, by friction with some coarse substance. Friction, it should be observed, is, either with or without bathing, an admirable substitute for exercise, when the latter cannot be conveniently taken. Although it is not attended with all the advantages derived from exercise in the open air, it nevertheless invigorates the muscular fibre, through sympathy with the cutaneous vessels, and increases the energy of the whole system.

The application of cold water has been recommended by many, as safe in all seasons and conditions. An indiscriminate practice upon such recommendation has often resulted in injury. During infancy, an extreme susceptibility to cold exists, and the system cannot readily resist the influence of a sudden change. The consequence of a powerful impression of this kind upon the sensitive membrane which encloses all the delicate vital organs is the rapid repulsion of blood towards them from the surface; and internal inflammation, by direct sympathy, is induced, which may prove destructive to life. The temperature of water should be *gradually* reduced, when applied to the young, as well as when used by the feeble and the aged. It must be modified to suit particular cases, both when adopted for the prevention and for the cure of disease.

The direct benefit of cold water, applied, as a curative agent, to the skin when preternaturally hot and dry, consists in the immediate reduction of temperature; and the diminished local action is communicated, by sympathy, to every portion of the cutaneous system of vessels; the balance of circulation thus being restored, and general inflammation dispelled. It is through this power and extent of sympathy that a tendency to certain inflammatory affections of the head and throat is counteracted by the application of cold to the feet; the temperature of the water being gradually reduced, when there is reason to apprehend a deficient reaction. It may be added, that, under the above-mentioned restriction, the practice of frequent bathing in this manner lessens the susceptibility to those sudden variations of temperature to which the feet, during nearly the entire year, are subjected. The consequences of wearing thin shoes might be rendered less fatal, were the feet more generally prepared, by the daily action of a cold bath, for the sudden and very great changes to which they are exposed by this pernicious practice.

The most convenient mode of ablution is that which consists in the use of a wet sponge or towel, the surface of the body being rubbed briskly with a dry cloth directly after the cold application. The habit of daily bathing in this way, the temperature of the water being regulated by the bather's sensations, will not only be found the best measure of prevention against those affections incident to our variable climate, but greatly promotive of strength, cheerfulness, and longevity.

We quote the following observations of Dr. Weiss, of England, in relation to this subject : —

" Ablutions may be local or general. They are performed in the following manner: The naked hands, or, better, a large sponge or woollen cloth, is dipped into a vessel containing cold water, placed upon a chair. The sponge is to be gently expressed, and then conveyed for some few minutes rapidly over the whole surface of the body ; water may also, at the same time, be poured over the head ; but not every one is able to bear the latter application, especially in the winter. Another method of performing ablution with cold water, consists in wrapping a linen sheet, dipped into cold water round the body, and thus washing the whole surface ; this process is more powerful, abstracting more heat from the body. In pursuing this or any other mode of ablution, it is advisable to stand in

a spacious vessel, so that the water which runs off
may not wet the room in which the operation is conducted.

"The best time, undoubtedly, for these ablutions is
the morning. They are to be performed immediately
on rising from bed, when the temperature of the body
is raised by the heat of the bed. The sudden change
favors, in a great measure, the reaction which ensues,
and excites the skin, rendered more sensitive by the
perspiration during the night, to renewed activity. In
some cases, and under certain conditions, more than
one of these ablutions becomes necessary; the same
operation may then be repeated at different intervals.
In most cases a second ablution, before going to bed,
will suffice. Local ablutions will have to be repeated
most frequently where we wish to produce increased
reaction; even in these cases, the temperature of the
body, or its natural warmth, should be restored before
proceeding to a second ablution. To increase the
beneficial effects of this washing, it should be accompanied by friction during the process; this is also
essential immediately after it. Quite as necessary is
exercise in the open air, if circumstances will in any
way permit it. Very great invalids only may be allowed, after washing, to retire to bed.

"Ablutions are, for the most part, preparations for a
more powerful system of treatment, in order to accus-
tom the body, by degrees, to water which is absolutely
cold. Tepid ablutions are, therefore, to be recom-

mended at first, especially to irritable and weakly individuals, or such as have never brought cold water in contact with their bodies.

"Ablutions continued for a quarter of an hour, or longer, act as a stimulant and refrigerant; those of shorter duration have a strengthening and exhilarating effect, and also the property of equalizing the circulation of blood, as may readily be perceived after general ablution of the whole body.

"The room in which ablution is performed may be slightly heated for debilitated patients in winter, to prevent colds in consequence of too low a temperature of the apartment; this exception is, however, only admissible for very weakly persons. Generally speaking ablution may be performed in a cold room, especially where persons get through the operation quickly, and can immediately afterwards take exercise in the open air."

ON AIR.

The relation which man sustains to the invisible element by which he is surrounded is of the most intimate nature. In the necessary act of respiration continually taking place, the lungs are filled with air; and it is estimated, that, during twenty-four hours, a quantity sufficient to fill fifty-seven hogsheads is inhaled by a healthy individual. It is highly important that this air should be pure; yet there are many circumstances that render it otherwise, and which will be alluded to in this chapter.

The atmosphere is composed of oxygen, nitrogen, and carbonic acid; the latter, however, existing in amount so inconsiderable as to be regarded rather as accidental than as forming one of its component parts. Oxygen constitutes a fifth part of air, and is the only ingredient that supports life; being, however, unfit for respiration unless combined with nitrogen in the proportion furnished by nature. In the process of breathing, a certain portion of the oxygen inhaled unites, in the lungs, with the carbon of the blood, for the purpose of renovating this fluid; while the remaining portion is exhaled, together with the nitrogen which has undergone very little alteration, and a quantity of carbonic acid gas, the result of the union of oxygen and carbon. This carbonic acid gas, when existing in larger quantity than is ordinarily contained in the atmosphere, is not suitable for respiration. The amount that proceeds from the lungs, in place of the oxygen retained, would not be injurious, were it allowed to pass freely into unconfined air; but, when breathed into a small, close space, it vitiates the air, and renders it wholly unsuited for further use. By excluding, therefore, pure air from rooms which one or more persons inhabit, a poisonous property accumulates, that more or less affects the health. Even though no immediate inconvenience is experienced, it is very evident that the air of a small apartment in which human beings reside during winter must

unavoidably soon become impure; for in such a situation, unless artificially ventilated in the most thorough
manner, there is not a sufficient supply of oxygen
introduced to replace that which is consumed. While
the vitiation of air in dwelling houses is not usually
so perceptible as to excite attention, yet a neglect of
proper ventilation is the cause of considerable illness.
In large establishments especially, where means of
ventilation are deficient, those diseases are very common which are believed to be directly produced or
aggravated by impure air. Scrofula is prevalent
among such as are subject to confinement in crowded
dwellings. Scurvy is said to be much more obstinate
among seamen in vessels that are not provided with
suitable ventilation. The mortality in hospitals has
been found to be very much decreased since means
have been adopted to purify the wards by the only
healthy process of introducing fresh air. Sir James
Clarke asserts that " living in an impure atmosphere
is even more influential in deteriorating health than
defective food, and that the immense mortality among
children and in workhouses is ascribable more to the
former than to any other cause."

The deadly effects of carbonic acid gas, when concentrated in wells and caves, and when disengaged
from burning charcoal, are well known. The result
is generally death, either immediate or after a short
interval, even when respiration of the carbonic acid
has been going on for a few moments only. This very
noxious gas is exhaled from the human lungs, as well

as from all the vegetable substances that are continually being mixed with the atmosphere; and while we
are certain that a large quantity of this poison is
capable of destroying life rapidly, it is to be fairly
inferred that the same deleterious property, though
in a less concentrated state, must, in confined situations, necessarily exercise an injurious influence upon
human health. In certain localities the atmosphere
becomes greatly vitiated by the decomposition of vegetable substances. The poison thus communicated to
the air has not the power of acting injuriously during
the day, since the air, being rarified by the rays of the
sun, ascends high above the surface of the earth; but
after sunset and before sunrise, the mist and dews precipitate these impurities; hence it is, that in the morning and evening the system suffers more from exposure
than at any other period. The moisture that rises
from low, swampy countries holds in solution poisonous
emanations or miasmata, and renders a residence in
such localities peculiarly unhealthy. In the regions
near Rome are extensive marshes, tainting the atmosphere for several miles around with a deadly effluvium,
so that travellers cannot ride over these districts during
the night without great danger of contracting fever.
The miserable residents there confined present a melancholy appearance, while wandering about gloomily,
with sallow complexion, sunken eyes, dropsical bodies,
and almost imbecile minds. There are many less
extensive but similar grounds in the valley of the
Mississippi. Heat attenuates these miasmatic exhala-

tions, causing them to be comparatively harmless. On this account, inhabitants near such prolific sources of sickness and death do much to avoid danger by con fining themselves, from sunset to sunrise, to rooms in which fires are lighted.

Animal matter in a state of putrefaction, neglected drains, collections of damp, decaying substances in cellars, back yards, and narrow lanes, are other prominent causes of deteriorated air, especially in crowded cities. Formerly, when no adequate measures were adopted for the removal of such refuse, diseases of the most serious nature were fearfully prevalent. Even at the present day, notwithstanding very excellent sanitary regulations, there still exist, among large collections of men, many causes of unhealthiness; one of which, and not the least conspicuous, is the unaccountably reckless practice of burying the dead near the homes of the living. There is yet much care to be exercised by municipal authorities to counteract the pernicious consequences arising from the crowding together of multitudes of human beings.

Miasmatic emanations, the product of decaying vegetation in damp localities, give rise to endemics, such as remittent, intermittent fevers, &c., that prevail in some portions of the United States and other countries; and there is no reason for doubting, that epidemics, like the plague and the cholera, are made much more virulent by, if not actually originating from, the noxious gases allowed to vitiate the air around human dwellings. Want of attention to this subject is new known to

27

have been followed by those terrible plagues that devastated European cities during the middle ages. Old drains, many of which remain in a neglected state for years in cities, generate sulphuretted hydrogen, a gas which is now known to produce the sickness that prevails on portions of the eastern coast of Africa; and which, if combined with common air in the proportion of a fifteen hundredth part of the gas, will destroy the life of birds, and a two hundredth part of the same will kill a horse.

At no period of life are the deleterious consequences of confinement in impure air more perceptible than during infancy and childhood. "Pale countenances, weak eyes, general relaxation of the body, an accumulation of all the inconveniences and sufferings of childhood, at length consumption and early dissolution of life, are the natural and frequent consequences of such confinement. The daily enjoyment of fresh air contributes greatly to the health and sprightliness of children, and is one of the most efficient preventives against that delicate and sickly condition which is so frequently witnessed in those who are almost constantly confined and pampered in nurseries."

ON EXERCISE.

By whatever circumstances man may be surrounded, however favorable his situation as to food, clothing, residence, freedom from hereditary disease, &c., his health cannot be preserved without exercise. The development of the muscular and osseous systems, of nervous and mental power, the vigorous circulation of the blood, the proper action of digestion, secretion, and nutrition, are dependent on a certain amount of physical exertion. The entire animal organization, as well as the constitution of external nature, prove that man was destined for an active existence. Through want of exercise, the supply of fresh material required to make up the loss of substance constantly effected by absorption, is interrupted; and the whole body is thereby deprived of firmness and solidity, becoming weak, languid, and emaciated. That this balance of waste and renovation may be maintained throughout the system, it is necessary that, in connection with pure air and suitable nourishment, the stimulus of exercise should be communicated to the muscles. By their properly regulated action, respiration becomes more full and free, thus purifying the blood of its carbon as fast as required; the heart's action is more rapid, thus hastening the circulation, and giving energy to all the processes whose office is to excrete useless, worn-out, unhealthy material.

An important consequence of well-ordered physical activity is the sense of comfort and cheerfulness imparted to the mind. There can be no positive happiness, aside from an ever-present consciousness of moral rectitude, but in an active state of the various systems constituting our bodies. The latter receive daily a definite amount of nervous energy, the expenditure of which, by means of muscular exercise, is attended with enjoyment. To the inactive, existence cannot prove a source of delight, physically considered; for various uncomfortable sensations are continually tormenting them, while living in neglect of that degree of exercise demanded by nature. It is a fortunate circumstance for him who · is forced to earn his subsistence by manual labor, that employment is a necessary condition of happiness. "God has dealt kindly with the great multitude," — the laboring class, — who are made happy by the very lot which may appear at times the least conducive to cheerfulness and contentment. It is a universal law throughout the human creation, that health is to be maintained only "by the sweat of the brow;" and for the existence of such a law the poor man may well be grateful.

With respect to the most beneficial kinds of exercise, the preference is to be given to walking; for in this mode of bodily action, especially if the arms are allowed to move freely, a more equal distribution of motion is imparted to the muscles in general, and the position is more unconstrained and natural than in any other. It is an exercise in which the rich and the poor man can, and should, at all times, indulge, and for the neglect of which no one possessed of sound limbs can offer an excuse. Whatever may be the individual's occupation, or the state of the weather, a daily walk of two hours, at least, should be taken. Much, however, of the benefit of exercise is lost, unless conducted in pure air. The circulation of the blood is accelerated by muscular exertion, and the motion of the lungs is consequently increased; but, for the proper oxygenation of the augmented flow of the blood to the lungs, fresh air is needed; and the more rapid the respiration, the more necessity exists for the inhalation of pure air.

Riding on horseback is a healthy mode of exercise, as it involves general muscular action, and is accompanied usually by the pleasurable excitement of companionship, with change of air and scene; but, under conditions of debility, it is too violent in its nature, unless the motion of the animal is easy, or the rider well instructed.

Dancing is, under certain circumstances, conducive to health, since the muscles are brought into united action, and the presence of company promotes the interest that should be felt in all exercise; but as conducted with us, in heated ball rooms, where the air is impure, and in dress which affords no protection against sudden changes of temperature, it is followed by more injury than benefit. Too many melancholy scenes of suffering and death have been directly traced to colds contracted in dancing assemblies.

The athletic exercises of the gymnasium, such as jumping, climbing, lifting weights, &c., if practised with moderation and judgment, contribute to the preservation of health. It is proper to add, however,

27 *

that, with those who have not gradually accustomed themselves to great exertion, an irregularity of action, from sudden spasmodic efforts, may lead to serious organic mischief.

Exercise, to be beneficial, should not be taken directly after a meal. It should be gentle and regular, never carried to the point of exhaustion, and always conducted, if possible, in the open air.

MATERIA MEDICA.

ACONITUM NAPELLUS.—ACONITE.—ACON.
(*Monkshood. Wolfsbane.*)

General Symptoms. The symptoms that lead to the selection of this medicine are, attacks of pain in head, back, and limbs, with thirst, hard and quick pulse, burning heat of skin, dry tongue and mouth, uneasiness, and want of appetite, with constipation — all constituting the common manifestations of fever; dizziness on sitting up; headache on moving the eyes; short, painful respiration; heavy and bruised sensation in the limbs; chills; scanty, difficult urination; pain and uneasiness aggravated at night; sensitiveness of the body to motion or contact; uneasy, dreaming sleep; dread of light; bitter taste in mouth; sense of oppression about the chest and stomach; shooting pains in chest, and in the region of the heart; lassitude.

ANTIMONIUM TARTARIZATUM.—ANTIM. T.
(*Tartrate of Antimony.*)

General Symptoms. Continued nausea; severe vomiting; sense of pressure in the stomach and abdomen; accumulation of phlegm in the throat; cough, with vomiting of food or mucus; difficulty of breathing;

paralysis of the lungs; trembling of the limbs, constant drowsiness; debility; aversion to food; paroxysms of suffocating cough; pustular eruption on skin; diarrhœa.

ARNICA (MONTANA).— ARN.
(*Mountain Arnica, Leopard's Bane.*)

General Symptoms. Convulsive movements in consequence of wounds; pain from bruises; pain in the teeth, with swelling of the gums; shooting pain in the stomach, extending to the back, and aching contraction of chest; expectoration of dark blood; painful weakness of the joints; congestion in head, with coldness of the body; fainting after external injury; hard, red swelling like biles; rheumatic and gouty pains; shooting pain in the chest and side.

ARSENICUM ALBUM.— ARSEN.
(*White Arsenic.*)

General Symptoms. Burning pains, internal; periodical attacks of pain; extreme weakness; emaciation; stiffness of the limbs; burning eruption on skin; burning under the skin, as if the blood vessels were filled with boiling water; intermittent febrile attacks; dropsical affections; irregular, weak pulse; inconsolable anxiety; throbbing headache on one side, attended with great sadness; cold perspiration, and frequent at night; inextinguishable thirst; asthmatic breathing; coldness of the surface; burning and sense of contraction in the throat; diarrhœa, with great debility; dry and dark tongue; bleeding and aphthous gums; ulcers, with a sensation of burning.

BELLADONNA (ATROPA).—BELL.
(*Deadly Nightshade.*)

General Symptoms. Fulness, weight and violent pain in the head, with heat, redness and swelling of the face, throbbing and fulness of the arteries of the neck; burning thirst; strong, full pulse; scarlet-red skin; daily headaches, increasing at night; swelling of parotid and sub-maxillary glands; soreness of throat; shooting pain in throat extending to the ears; suffocating cough; hoarseness; inflammation of the tongue; cramp, spasms, and convulsive movements of the limbs; erysipelatous inflammation; red, shining swellings; sleeplessness; aching pain in eyes, extending into the head; rapid and anxious respiration; painful swelling and stiffness of the neck; cramp-like pains, and soreness of the abdomen; swelling and scarlet redness of the arms and hands; heaviness and numbness of the lower limbs; shooting pain in the loins.

BRYONIA (ALBA).—BRY.
(*White Bryony.*)

General Symptoms. Drawing pains, especially in limbs, increased by motion or by contact; giddiness on rising; headache, with giddiness, followed by chills; burning, aching, or pulsative pain in the stomach after eating; breathing impeded by darting pains in the chest and side, that are increased by motion; palpitation of the heart; spasmodic cough, especially after eating, and often with vomiting; morbid appetite; pain, with shivering and coldness of the body; internal

heat; pain attended with great irascibility and dis-
couragement; jerking pains in the teeth; buzzing in
the ears; regurgitation of food after a meal; cough,
with yellowish or bloody expectoration; swelling of
the feet, with heat and redness; all pains increased by
moving; constipation.

CALCAREA CARBONICA.—CALC.
(Carbonate of Lime.)

General Symptoms. Cramps in the limbs; excessive
agitation; fatigue from speaking or walking; sensibility to cold, damp air; debility; icy coldness in and
about the head; profuse perspiration after moderate
exercise; painful sensibility of teeth and gums; bitter,
metallic taste, particularly in the morning; sensation
of hunger after eating; morbid appetite; swelling of
the upper lip; long-continued hoarseness; cough, with
purulent expectoration; shooting pains in chest and
side; stiffness and weight of lower limbs; bloated condition of the body; difficult dentition; swelling of the
knee joints; weakness and numbness of the fingers;
indurated swelling of the glands; diseases of a scrofulous nature; premature menstruation; curvature of the
spine.

CAMPHORA.—CAMPH.
(Camphor.)

General Symptoms. Convulsions and cramps; coldness over the body, with blueness of the skin; loss of
consciousness; burning heat in the stomach and abdomen; cramps in the lower limbs; spasms in the chest,

with suffocating oppression; giddiness, and weight of head; epileptic fits.

CANNABIS.—SATIVA.
(*Hemp.*)

General Symptoms. Obstinate urinary retention; burning pain during urination; difficult respiration, with palpitation of the heart; general dejection; fatigue after slight exertion, with trembling of the limbs.

CANTHARIS.—CANTH.
(*Spanish Fly.*)

General Symptoms. Difficult urination; retention; cutting pains during urgent and ineffectual efforts to overcome the retention; inflammation of the kidneys; prostration of strength.

CARBO VEGETABILIS.—CARB. VEG.
(*Vegetable Charcoal.*)

General Symptoms. Bruised feeling or numbness of limbs, especially in the morning; extreme prostration; liability to take cold easily; throbbing weight in head, especially in the evening; difficulty of breathing, with aching pain in chest; nocturnal perspiration; extremely weak pulse; headache from being overheated; retraction, ulceration, and bleeding of the gums; acid taste in mouth, with inflation of the abdomen after eating; flatulent colic; great dejection with debility; pressure and burning pain in the eyes; pain in the back from lifting heavy weights; great accumulation of phlegm in throat; emaciation; disturbances in stomach and bowels after eating.

CHAMOMILLA (VULGARIS).—CHAM.
(*Chamomile.*)

General Symptoms. Great nervous irritability, and sensibility to pain; drawing pains at night, with thirst, heat, and redness of one side of the face especially; attacks of convulsions and spasms; extreme agitation and sadness; irritability; earache, and toothache, chiefly on one side, increasing at night, by warmth; bitter taste of food; want of appetite; excessive thirst for cold drinks; bilious vomiting; cutting pains in abdomen; dry cough, with constant tickling in throat; disposition to sadness; the evil effects of anger or grief; diarrhœa, particularly in children; menstrual colic; difficult breathing, with sense of fulness and stinging sensation in chest; extremely painful pressure in the region of the heart.

CINCHONA (OFFICINALIS).—CINCH.
(*Peruvian Bark.*)

General Symptoms. Great general weakness, with trembling and tendency to perspire on the least motion, and during sleep; over excitability of the nervous system; dropsical swellings; tearing pain in the limbs excited or aggravated by touch or movement; attacks of intermittent fever; pressing headache, and great sensibility of the head to touch; bleeding of the nose; throbbing toothache; weakness of digestion; acid eructations; diarrhœa; convulsive cough, with shooting pains in chest; pains in the side, with palpitation of the heart; numbness and spasmodic action of the upper and lower limbs, especially when touched; weakness and swelling of the joints.

CINÆ SEMEN. — CINA.
(*Mugwort of Judea.*)

General Symptoms. Dull pains in the head, with dilated pupils; paleness of the face, with livid circle under the eyes; irritation of the nose; craving for food; short, dry cough; rapid respiration; worms.

COCCULUS INDICUS. — COCC.
(*Indian Berry.*)

General Symptoms. Semi-lateral sufferings; convulsive movements of the limbs, as in St. Vitus's dance, want of vital energy; vertigo, as from intoxication, with nausea and vomiting; dull headache; flatulent, cramp-like colic; premature menstruation, with cramps in abdomen; constipation; short, interrupted respiration.

COFFEA (CRUDA).
(*Raw Coffee.*)

General Symptoms. Excessive nervous excitability; neuralgic pains; evil consequences of excessive joy; sleeplessness, from nervous excitement; semi-lateral headache; sharp pains in teeth or throat, with extreme sensibility; short, interrupted breathing; deafness, with a humming sound in the ears; diarrhœa, during dentition.

COLCHICUM (AUTUMNALE). — COLCH.
(*Meadow Saffron.*)

General Symptoms. Nervous fatigue from long watching; rheumatic pains in the limbs, with numb

ness ; tearing pains in teeth ; insipidity of food ; nausea ; sensation of coldness in the stomach ; pain in the
abdomen ; dysenteric diarrhœa.

COLOCYNTHIS. — COLOC.
(*Bitter Cucumber.*)

General Symptoms. Evil effects of excessive emotion, of indignation, or mortification ; excessively violent, spasmodic colic ; painful cramps and cramp-like
contractions ; colic and diarrhœa, however little may
be eaten ; semi-lateral sufferings ; dysentery, attended
with colic pains ; restlessness ; mental dejection.

CUPRUM METALLICUM. — CUPR.
(*Copper.*)

General Symptoms. Violent vomiting ; cramps in
the abdomen, diarrhœa, and convulsions ; exceedingly
troublesome sensation of pressure in stomach and bowels, aggravated by touch and by motion ; spasmodic
colic, with convulsions ; cramps in the chest, with dry
cough and accelerated respiration ; epileptic fits, followed by chills and headache ; spasms in fingers and
feet.

DIGITALIS (PURPUREA). — DIG.
(*Foxglove.*)

General Symptoms. Dropsical swellings ; extreme
debility, even to fainting ; pain in the region of the
heart, with slow pulse, and blueness of skin ; want of
appetite ; giddiness ; nausea, with aching weight and
fulness in the stomach ; sensitiveness in the region of
the liver ; difficult breathing ; dry cough.

DROSERA (ROTUNDIFOLIA). — DROS.
(*Sundew.*)

General Symptoms. Dry, spasmodic cough, with in-clination to vomit; vomiting of food during cough, and afterwards; cough, with purulent expectoration; hoarseness; fatiguing cough, with bluish color of the face, whistling respiration, bleeding of nose and mouth, with apparent danger of suffocation; loss of sight from excess of light; chills, with coldness of feet, hands, and face.

DULCAMARA (SOLANUM). — DULC.
(*Bitter-Sweet.*)

General Symptoms. Affections resulting from taking cold; shooting pains, aggravated at night; dropsical swellings from cold; lassitude; sore throat from a cold; diarrhœa from cold; colic pain from a cold; dry heat and burning sensation in the skin; cough, whooping, excited by inspiring deeply; eruption on skin, like tet-ters; extreme desire for cold drinks; nettle rash; stu-pefying headache, increased by the slightest motion, or by speaking; eruption, like warts, on the face; swelling of sub-maxillary glands.

EUPHRASIA (OFFICINALIS). — EUPHR.
(*Eyebright.*)

General Symptoms. Inflammation and ulceration of the eyes, and of the eyelids; abundant flow of tears; cough in the morning, with flowing coryza and abun dant expectoration; spasmodic pains in the back and limbs.

HELLEBORUS (NIGER). — HELL.
(*Christmas Rose.*)

General Symptoms. Dropsical swellings ; painful heaviness of the head, and great sensibility of the exterior of the head, especially at the back part, as if bruised ; disposition to bury the head in the pillow when sleeping ; heaviness and fulness of the stomach, with extreme uneasiness ; difficult respiration.

HEPAR SULPHURIS. — HEP.
(*Sulphuret of Lime.*)

General Symptoms. Barking cough ; attacks of dry, rough, and hollow cough, with distress and apparent danger of suffocation, especially in the evening ; scraping sensation in the throat, that obstructs speech and deglutition ; whistling respiration ; great flow of saliva ; unhealthy skin, with chaps and ulcerations ; inflamed eyes ; purulent discharge from the ears ; suppurations ; swelling of the tonsils.

HYOSCYAMUS (NIGER). — HYOS.
(*Henbane.*)

General Symptoms. Excessive nervous excitement, with sleeplessness ; convulsive movements ; epileptic symptoms, as unconsciousness, foaming at the mouth, blueness and swelling of the face, protruded eyes ; attacks of cerebral congestion; vertigo, with benumbing pain in the forehead ; balancing of the head from side to side ; delirium ; eyes fixed or convulsed ; pulsating pains in teeth, with heat and redness of the face ; spasmodic cough at night, with redness of face and mucous vomiting; also dry, shaking cough, with soreness of the abdominal muscles; fever, with extreme debility.

IGNATIA (AMARA). — IGN.
(*St. Ignatius's Bean.*)

General Symptoms. Evil effects of grief or vexation; spasmodic affections, resulting from fright or contradiction; alternate states of gayety and sadness; hysterical debility, and fainting fits; universal heat, with internal chills; pain in head and stomach, with sensation of great fatigue; want of appetite and want of thirst; nausea and vomiting of food; uterine cramps, with painful and too frequent menstruation; sensation of emptiness and weakness in the stomach; insupportable pains in the bones and joints of the limbs.

IPECACUANHA. — IPEC.

General Symptoms. Disgust for all food; attacks of sick headache; nausea and vomiting of undigested food or of bile; dysenteric diarrhœa, attended with nausea and colic; dry, spasmodic cough; great and sudden weakness; short, anxious breathing; spasmodic asthma, with contraction of larynx; loss of breath on the least motion; hemorrhage from various organs; bad effects from the use of bark or quinine.

LACHESIS (TRIGONOCEPHALUS). — LACH.

General Symptoms. Intermittent, periodical sufferings; great mental and bodily prostration; attacks of convulsions, with epileptic symptoms; deep, pressing pains in the head; difficulty of swallowing and of

28 *

breathing; pain in throat, aggravated by the touch; hoarseness; pain in the stomach, attended with eructations, nausea, sensitiveness; constipation; palpitation of the heart; toothache, with pain in the head; chills, heat, and weight in limbs; vertigo in the morning; dry, short cough; hemorrhoids, with colic; pain and swelling in abdomen; aggravation and renewal of troubles after sleep.

LEDUM (PALUSTRE). — LED.
(*Wild Rosemary.*)

General Symptoms. Rheumatic and gouty pains, aggravated by motion, in the evening, and by warmth; inflamed swelling of the knees or feet, with stinging, tearing pains; dry, tettery eruption on the face; cough, with expectoration of bright red blood.

LYCOPODIUM. — LYC.
(*Clubmoss.*)

General Symptoms. Rheumatic pain in limbs and joints, with stiffness and numbness; continued nausea, with aching and swelling of stomach; constant aching or shooting pain in the chest; obstinate, dry cough, with short breathing, smarting in chest, and pain in stomach; constant oppression in chest, increased by fresh air; shooting pains in the back; constipation; diseases of bones, with increase of pains at night; swelling of the glands of the neck; white swelling of the knee and foot; gouty stiffness in the elbow and wrist; giddiness on stooping; tearing pain in forehead, chiefly towards night; slow fever, with clammy perspiration.

MERCURIUS. – MERC.
(*Mercury.*)

General Symptoms. Severe rending pains at night; rheumatic pains, with profuse perspiration that affords no relief; excitability; enlargement, inflammation, and ulceration of the glands; easily bleeding eruption, resembling " scabies; " pains in the bones, as if they were broken ; profuse perspiration in fevers; livid, sensitive gums, and shooting pains about the roots of the teeth ; inflammation and ulceration of mouth and throat; dryness, and shooting pain in throat; pain in the eyes, with deafness ; nausea ; weight and tenderness in stomach and liver ; excessive pains in abdomen, subsiding on lying down ; dry, convulsive cough, accumulation of phlegm in throat; catarrh ; dysentery.

NUX VOMICA. – NUX VOM.

General Symptoms. Sufferings from the use of coffee, stimulants and narcotics ; from passion, study, watching, sedentary habits ; shooting pains in, with stiffness, weakness, and numbness of limbs and joints; great prostration in the morning ; nervous excitement; fits of uneasiness, with nausea, anxiety and trembling: pains arising within doors disappear on going out, and vice versa ; aggravated, or first felt in the morning ; intermittent fever, with morning exacerbations, with congestive headache and nausea ; feverish chills, with giddiness, headache, and thirst ; pain, as if a nail were driven into the brain; heaviness in the head, with a sensation of expansion ; dry cough, with headache and catarrhal symptoms, increased by moving; piercing pain in the teeth, extending to the ears and cheek bones, renewed by drinking; nausea and vomiting; weight and burning pain in stomach, chiefly after eating, and in the morning; spasmodic pain and swelling in abdomen; stricture in chest, and pressure, with asthmatic breathing; obstinate constipation; pains in the back.

OPIUM. — OP.

General Symptoms. Convulsive movements and spasms, with epileptic symptoms; muscular rigidity; extreme desire, with entire inability to sleep; vertigo and stupefaction; congestion of the brain; twitching and relaxation of the muscles of the face; delirium, with frightful visions and trembling; fixed or convulsed eyes, with dilated pupils; long-continued constipation; difficult and noisy respiration.

PETROLEUM. — PETR.
(*Rock Oil.*) .

General Symptoms. Vertigo; great weakness after the least effort; nausea after eating, and when riding or sailing; swelling and sense of fulness of stomach; throbbing pain, with weight and fulness in head; swelling of the gland sof the neck; itching and burning pustular eruption; chaps; chilblains.

PHOSPHORUS. — PHOS.

General Symptoms. Gouty and rheumatic rending and shooting pains in the limbs after slight exposure to cold; nervous debility; frequent giddiness, with congestive headache; cough, with purulent expectoration.

or of adhesive mucus or blood; painful sensibility of the larynx; hoarseness, scraping sensation in the throat; loss of voice; difficulty of breathing, and oppression in the chest; lancinating pain in left side of chest; sensation of burning and excoriation in chest, with palpitation of the heart, and a daily dry cough; chronic, prostrating diarrhœa; stiffness of knee joint, with swelling and numbness of the feet.

PHOSPHORIC ACID. — PHOS AC.

General Symptoms. Great general weakness; sufferings from loss of the fluids; bad consequences of vexation and anxiety; typhoid fevers; night sweats; swelling of the bones; eruption of pimples on the face; scarlet eruption on different parts of the body; diarrhœa, during the prevalence of epidemic cholera.

PULSATILLA. — PULS.
(*Wind Flower.*)

General Symptoms. Evil consequences from the use of sulphur, mercury, quinine, oily food, as the fat of pork, &c.; from fright, or a chill in the water; rheumatic and gouty pains, worse at night, and pain shifting rapidly from place to place; measles; periodical and semi-lateral affections; shooting, throbbing, one-sided or dull pain in the head, with dizziness after lying down towards night; headache from fat, indigestible food; jerking toothache, extending to the ears and head, lessened by cold, and increased by warmth; all sufferings aggravated towards night; nausea, vomiting, and pressure in the stomach after eating; painful sen sitiveness and cramp in stomach; colic pains, often attended with vomiting and diarrhœa; hemorrhoids; uterine spasms; suppressed and generally disturbed menstruation; leucorrhœa; catarrh, hoarseness, shaking cough, with expectoration; nausea, palpitation, shooting pain in chest; pain in lower part of back; pain, weariness, and swelling in the lower limbs.

RHUS TOXICODENDRON. — RHUS T.
(*Poison Oak.*)

General Symptoms. Shooting, wrenching pain in the joints, with stiffness and swelling of limbs; pain, as if the flesh were torn from the bone; pains increased by rest, and alleviated by motion; itching, burning vesicles on a red base; erysipelatous, vesicular eruption, with swelling; malignant fever, attended with extreme weakness, violent pain in the limbs, and delirium; periodical headaches, extending to the ears, cheek bones, and teeth; eruption on the head, with incrustations that destroy the hair; inflammation and swelling of the parotid glands, known as "mumps;" constipation, alternating with diarrhœa; pains in the back, as if strained by lifting; erysipelas on limbs; cough, with expectoration of bright red blood.

SAMBUCUS (NIGRA). — SAMB.
(*Elder Tree.*)

General Symptoms. Spasmodic asthma and cramps in the chest; fever, with profuse perspiration at night; attacks of suffocating cough, with cries, in children; wheezing respiration; spasmodic paroxysm of suffocatiun at night, with blue and bloated face ; catarrh in children.

SEPIA. — SEP.
(Juice of Cuttle Fish.)

General Symptoms. Vertigo and headache, with nausea and vomiting ; expansive pressure in the head when stooping ; uneasiness and throbbing in all the limbs ; great tendency to take cold ; herpetic eruption on face ; inflammation of the eyes and eyelids ; burning, shooting pain, and oppression in stomach, after taking food, and while walking ; corrosive leucorrhœa ; painful or suppressed menstruation ; cough, with much expectoration, salt, mucous or bloody, sometimes with nausea and vomiting ; shortness of breath, and oppression while walking ; aching pain in chest, when moving, or at night ; rheumatic pains, with pustular or ulcerated eruptions on different parts of the body.

SILICA. — SIL.
(Silex.)

General Symptoms. Inflammation, induration, and ulceration of the glands ; diseases of the bones ; ulcers of various kinds ; chronic coryza ; great tendency to suffer from cold, with great nervous debility ; herpetic and scabious eruption on face ; scrofulous affections ; deafness ; chronic cough, with hoarseness and purulent expectoration ; debilitating morning sweat ; abscesses ; mercurial poisoning ; eruptions on the head ; worms ; bruised and paralytic sensation in the limbs ; shooting pain in the teeth, at night, with easy bleeding of the gums ; loss of smell, and want of appetite, with water brash ; sufferings aggravated by motion or contact.

SPIGELIA. — SPIG.
(*Indian Pink.*)

General Symptoms. Gouty, shooting, and tearing pains in the limbs; great lassitude after slight exercise; inflammation of the eyes; pulsating toothache, after eating, aggravated by cold; lancinating pain in the region of the heart, with dry cough, impeded breathing, and spasmodic contraction of the chest.

SPONGIA TOSTA. — SPONG.
(*Burnt Sponge.*)

General Symptoms. Hollow, dry, barking cough; hoarseness and pain in the larynx; wheezing, respiration, with sense of obstruction in the throat; feverish heat, with dry skin, and thirst.

STANNUM. — STAN.
(*Tin.*)

General Symptoms. Attacks of epilepsy; excessive emaciation; debilitating night sweats; hoarseness and roughness in the throat; cough, with muco-purulent expectoration, excited by talking, or by lying on the left side; difficulty of breathing, with pain, and sensation of weakness in the chest; swelling of the feet; leucorrhœa, with great prostration of strength.

STRAMONIUM (DATURA). — STRAM.
(*Thorn Apple.*)

General Symptoms. Cramps and hysterical sufferings; convulsive movements; suppression of secretions generally; delirium, with frightful visions; mania;

stammering speech, and obstructed deglutition; trembling of the limbs.

SULPHUR.—SULPH.

General Symptoms. Sufferings of lymphatic temperaments; subject to cutaneous eruptions, swollen glands, hemorrhoids, hypochondria, liability to cold; inflammatory swelling of joints; spasms, convulsions, fainting fits; excessive fatigue after moving or talking; extreme emaciation; itching eruptions on skin, almost insupportable at night; vertigo, pressure and heaviness, or sharp pains in head; periodical headaches; coldness of the head; sharp pains in the ears; bleeding of the nose; sharp, throbbing pains in the teeth, with swelling of the gums; aphthous ulcerations in the mouth; pressure in throat as from a tumor; immoderate appetite; regurgitation of food; nausea, with oppression after a meal; shooting pain and induration in the region of the liver, and in the abdomen; heavy pressure in abdomen, as from a stone; sufferings aggravated at night, or when standing; obstinate constipation; hemorrhoids; premature menstruation, or entire suppression, with colic, spasms, headache, pain in the loins; leucorrhœa; cough, with profuse expectoration, and shooting pains in chest, short breath, and weakness; livid spots on skin from slight bruises; incarcerated hernia; erysipelatous eruptions on limbs; coldness of feet.

29

SULPHURIC ACID.—SULPH. AC.

General Symptoms. Small, red, livid spots as from a bruise; aphthæ in the mouth; inguinal hernia; chronic diarrhœa; cough, with expectoration of blood; difficulty of breathing; chronic inflammation of the eyes.

VERATRUM ALBUM.—VERAT.
(*White Hellebore.*)

General Symptoms. Violent and painful diarrhœa, with continued nausea and vomiting, excited by a drop of liquid or the least movement; insatiable thirst; strong desire for acid or cold food; great exhaustion, chills, with cold perspiration; cold hands and feet; severe cramps in lower limbs; internal heat, with delirium; intermittent fever; suppressed menstruation, with delirium; pain in the loins and back, as if beaten; voracious appetite; mental alienation; violent delirium

To which reference has frequently been made in this work, is the chief external application used in the practice of homœopathy. It is advantageously employed in cases of sprains, contusions from a blow or fall, shocks, and nearly all injuries from foreign substances to which the surface of the body is liable. The Arnica is to be mixed with water, in the proportion of one part of the tincture to four or five parts of water, and a linen cloth dipped in this mixture is to be laid upon the injured part, and renewed, from one to three or six hours, according to the severity of the injury. At the same time a dose of Arnica (five or six globules) should be administered as frequently as the external application is made, until there is a decided mitigation of the attendant pain, and the general irritation.

MANNER OF PREPARING THE TINCTURE OF ARNICA.

All parts of the Arnica Montana are used in medicine, but the flowers only are readily obtained here, and they are in many respects to be preferred. These, when dried, are of a yellow color, have an aromatic odor, and an acrid, disagreeable taste. The flowers are to be steeped for eight or ten days in alcohol, in the proportion of one ounce of the flowers to one pint of alcohol. The tincture is then fit for use, diluted as above directed.

339)

EXPLANATION

Absorbents. The small, delicate, transparent vessels which take up substances from the surface of the body, or from any cavity, and convey them to the blood, are termed absorbents, or absorbing vessels.

Acute. As applied to disease, signifies one that is manifested by violent symptoms, terminating in a few days, and attended with danger.

Adipsia. A want of thirst.

Adynamic. Deficient vital power.

Ætiology. The doctrine of the causes of disease.

Agglutination. The adhesive union of substances.

Aliment. Every substance or fluid which is capable of affording nourishment.

Alkali. This term is used to designate the property which certain substances possess of neutralizing acids, and by combining with them to form salts.

Aneurism. A pulsating tumor formed by the dilatation of an artery.

Angina. A sore throat, with difficulty of swallowing.

Anorexia. A want of appetite, without absolute loathing of food.

Anosmia. Loss of the sense of smell.

Antidote. A term used to denote a medicinal agent that has the power of modifying or neutralizing the effects of another medicine previously administered, or of counteracting poison.

Antiphlogistic. A term applied to those medicines, plans of diet, or other influences which tend to oppose inflammation, or which, in other words, weaken the system by diminishing the activity of the vital power.

Aphonia. A loss of voice.

Aromatics. A term applied to substances of an agreeable, spicy, or pungent taste, as cinnamon, cardamom seeds, &c. .

Arteries. Two membranous canals that arise from the right and left sides of the heart, and are distributed, gradually growing less as they proceed, to all parts of the body, conveying the blood that is to nourish, generate heat, and preserve life. They receive names from their different locations, and terminate either in the veins or the capillary vessels. The action of the arteries, called the " pulse," corresponds with that of the heart, and depends upon the impulse given to the blood by that organ.

Arthritic. Pertaining to the gout.

Asphyxia. The state of the body, during life, in which the pulsation of the heart and arteries cannot be perceived.

Assimulation. The conversion of food into nourishment.

Asthenia. A state of extreme debility.

29 *

Astringent. That which, when applied to the body, renders the solids denser and firmer by contracting their fibres, independently of their living, muscular power.

Atony. A weakness or deficiency of muscular power.

Attenuant. That which possesses the power of imparting to the blood a more thin or fluid consistence than it had previous to its exhibition, such as water, whey, and all aqueous fluids.

Balsamic. A term generally applied to substances of a smooth and oily consistence, which possess emollient, sweet, and aromatic qualities.

Bilious. A word used to denote diseases which arise from a too copious secretion of bile, as bilious colic, bilious fever, &c.

Brachial. Of or belonging to the arm.

Bronchotomy. An operation in which an opening into the windpipe is made for the purpose of introducing air into the lungs when any disease, as croup, for example, obstructs breathing by the mouth, or for the purpose of removing any substance which may have passed into the windpipe.

Buccal. Appertaining to the cheek.

Bulimia. Insatiable hunger.

Cachexia. A bad habit of body, known by a depraved or vitiated state of the solids and fluids.

Capillaries. The very small ramifications of the arteries which terminate upon the external surface of the body, or on the surface of internal cavities, **are** called capillary vessels, or capillaries.

Caries. A mortification of the bones.

Cartilage. A white, elastic substance connecting bones together, or to the joints.

Cellular Tissue. A fine, net-like membrane, composed of laminæ and fibres joined together, which is the connecting medium of every part of the body.

Cerebral. Relating to the brain.

Cervical. Belonging to the neck.

Chalybeate. Of or belonging to iron. Chalybeate springs are mineral waters abounding in iron.

Chronic. A term applied to diseases which are of long continuance, and mostly without fever. Used in contradistinction to the term " acute."

Chylification. The process carried on in the small intestines, by which the chyle is separated from chyme.

Chymification. The first stage of digestion, performed in the stomach on the food.

Coagulation. The separation of the coagulable particles contained in any fluid from the more thin and non-coagulable portion ; thus when milk curdles, the coagulable particles form the curd; and when acids are thrown into any fluid containing coagulable particles, they form what is called a " coagulum."

Coma. An inclination to sleep ; a lethargic drowsiness.

Coma Vigil. An inclination, but an inability, to sleep

Congestion. An unusual accumulation of blood in any particular set of vessels. On the brain, chest, &c., it denotes an over-distention of the vessels there situated.

Confluent.   Running together.   Applied to eruptions on the skin.

Contagion.   The application of any poisonous matter to the body through the medium of the touch.

Cubital.   Belonging to the forearm.

Cutaneous.   Belonging to the skin.

Cuticle.   The thin, insensible membrane that covers and defends the true skin.

Cutis.   The true skin, under the cuticle.

Desquamation.   The separation of laminæ or scales from the surface of the skin or from bone.

Diagnosis.   The science which delivers the signs by which one disease may be distinguished from another; the distinguishing symptoms are termed " diagnostic."

Dietetic.   Relating to diet.

Depurative.   Applied to those agents which free the system from impurities.

Demulcent.   Mild, viscid fluids which protect tender surfaces from the action of acrid or irritating substances.

Dysecoia.   Difficulty of hearing.

Dysphagia.   Difficulty of swallowing.

Dyspnœa.   Difficulty of breathing.

Dyplopia.   Double vision.

Ectropium.   An eversion of the eyelids.

Eclampsia.   A scintillation or flashing of light that is frequently perceived by epileptic persons.

Ecchymosis.   A black and blue swelling that arises from a bruise, or from spontaneous extravasation of blood.

Effusion. The escape of blood or other fluid from the vessels containing it.

Efflorescence. An unnatural redness of the skin.

Endemic. An affection peculiar to a certain class of persons, or to certain localities.

Epidemic. A contagious disease, which attacks several persons at the same season, and in the same place.

Exanthema. Cutaneous eruptions, rash-like.

Excretion. Applied to the removal from the blood of those fluids or substances that serve no useful purpose in the animal economy.

Exfoliation. The separation of a dead piece of bone from the living.

Extravasation. Applied to fluids which are out of their proper vessels or receptacles.

Facial. Belonging to the face.

Farinaceous. A term given to all articles of food which contain farina or flour.

Fauces. The cavity behind the tongue, at the entrance of the throat.

Flatulence. A great accumulation of air in the stomach and intestines.

Fomentation. A term applied to partial bathing, when wet cloths are laid upon any part of the surface, whereby perspiration is promoted.

Fumigation. The application of fumes to destroy contagious miasmata or effluvia. The most efficacious substance for this purpose is chlorine ; next to it the vapor of nitric acid ; and lastly that of muriatic acid.

Functional. Applied to diseases in which there is

merely a deranged action, without loss or injury of substance.

Ganglion. A natural, knot-like enlargement in the course of a nerve. Also applied to an unnatural tumor, formed in the sheath of a tendon, and containing a fluid. This tumor is situated on the back of the hand, or on the foot.

Gangrene. A loss of vitality of any portion of the body; synonymous with mortification.

Gastric. Appertaining to the stomach.

Globus Hystericus. The air rising in the throat, and prevented by spasm from reaching the mouth, producing the sensation of a round body in the throat, is so called because it is a symptom attending hysteria.

Globule. The little round grain, made of sugar of milk, and saturated with any remedy used in homœo pathic practice, for the more convenient administration as well as better preservation of medicine.

Granulation. The little, grain-like, fleshy bodies which form on the surface of ulcers and suppurating wounds, serving to unite the sides and fill up the cavities.

Hectic. Febrile action symptomatic of diseased lungs and other organs, with exacerbations or increase of fever twice every day, about noon and towards night.

Hemorrhage. A flow of blood.

Hepatic. Belonging to the liver.

Homœopathy. The cure of disease by medicines

that are capable of producing in a healthy person symptoms similar to those existing in the disease.

Hemiplegia. A paralytic affection of one side of the body.

Hemeralopia. A defect of sight in consequence of which the person sees only during the day, and not a night.

Hemiopia. A defect of sight where the person sees only one half of an object.

Idiopathic. A disease which does not depend on, and is in no way related to, any other complaint.

Idiosyncrasy. A peculiarity of constitution in which a person is affected by certain agents, which, if applied to a thousand other persons, would produce no effect.

Intermittent. When applied to the pulse, means an irregularity of pulsation ; the beating of the artery ceasing for two, three, or four seconds, and then returning to beat regularly for a short time, which regularity is succeeded by another interval of cessation, and so on. When applied to fevers, means an entire cessation of febrile symptoms for a certain period ; the paroxysms of fever coming on once or twice a day, every second or third day, leaving intervals of absence from all feverish indications.

Irritation. The action produced by any stimulant.

Infection. Applied to diseases that are produced by local causes contaminating the atmosphere, and not necessarily communicated by the contact of bodies.

Jugular. Belonging to the throat.

Lamina. A bone, or membrane, or any substance resembling a thin plate of metal.

Lethargy. A heavy sleep, with scarcely any intervals of waking.

Ligaments. Strong elastic membranes connecting
the extremities of movable bones.

Ligature. Waxed thread or silk of various thickness prepared for tying severed vessels, or uniting
incised wounds.

Lotion. A liquid application to the skin.

Luxation. The dislocation of a bone from its natural
cavity.

Lymphatics. Absorbent vessels that carry a transparent fluid called "lymph."

Malaria. An atmosphere infected by animal and
vegetable decomposition, that produces endemic or
local diseases.

Malignant. A term which may be applied to any
disease, the symptoms of which are so aggravated as to
threaten the extinction of life.

Maturation. A word denoting that process which
succeeds inflammation, by which pus is collected in an
abscess.

Medullary. The white or internal substance of the
brain.

Megrim. A species of headache, affecting one side,
in the region of the temple.

Mephitic. Having a disagreeable, noxious vapor.

Metastasis. The translation of a disease from one
place to another.

Mortification. The loss of vitality of any part of the
body.

Mucilage. An aqueous solution of gum.

Narcotics. Medicines which have the power of producing sleep.

Nosology. The doctrine of the names of diseases, or the arrangement of diseases into classes, orders, genera, &c.

Nostrums. Applied to all quack medicines, the composition of which is kept a secret from the public, and known only to the inventor.

Occipital. Belonging to the occiput, or back part of the head.

Olfactory. Belonging to the sense of smell.

Ophthalmic. Belonging to the sense of sight.

Opisthotonos. A spasm of the muscles by which the body is bent backwards and kept in a fixed position.

Organic. Of or belonging to an organ. This term is also used to distinguish a disease of structure from a functional disease. Thus, when the substance of the liver is changed by disease, it is called an *organic* disease; but when it merely furnishes vitiated bile, the disorder is said to be *functional.*

Orthopnœa Rapid and labored respiration, during which the person is obliged to preserve an erect posture.

Ossification. The process of conversion into bone.

Oxygenation. A union with oxygen.

Palliatives. Medicines which relieve without curing disease.

Papular. Used to designate an eruption or small elevation of the skin, with an inflamed base, not containing a fluid nor tending to suppuration.

Parenchyma. The cellular substance which connects parts together.

Paroxysm. An increase of the symptoms of a disease that continues for a longer or a shorter period, and then declines.

Pathogenetic. A term applied to the effects produced upon a healthy individual by taking any medicine.

Pathognomonic. A term given to those symptoms which are peculiar to a disease.

Pathology. The doctrine of diseases.

Pectoral. Of or belonging to the chest.

Pediluvium. A bath for the feet.

Periosteum. The membrane that invests the bones.

Petechiæ. Dark red or purple spots, resembling the bites of fleas.

Physiology. That science which has for its object the knowledge of the phenomena proper to living bodies.

Plethora. A superabundance of blood.

Prognosis. The foretelling the event of disease from particular symptoms.

Prophylactic. Any means made use of to preserve health and prevent disease.

Psora. A cutaneous eruption, characterized by continued itching. The founder of homœopathy regarded this disease as existing in most individuals in a latent state, and that this condition was peculiarly favorable for the production of diseases through the influence of injurious agencies.

Pulmonary. Belonging to the lungs.

Purulent. Having the appearance of pus.

Pus. A white, cream-like fluid, exuding from injured surfaces, or contained in an abscess.

Pustule. An elevation of the cuticle, globular or conoidal in form, and containing pus.

Pyrexial. Appertaining to fevers.

Regimen. A term employed in medicine to denote the plan or regulation of the diet.

Remittent. When feverish symptoms diminish considerably, but not entirely, at certain periods, to recur with their previous severity after the amelioration, they are called remittent.

Resolution. A termination of inflammation in which the disease disappears without suppuration or mortification being occasioned.

Retrocedent. When a disease that moves about from one part to another, and is sometimes fixed, has been for some time in its more common situation, and retires from it, it is said to be retrocedent.

Sanguification. A natural function of the body, by which the chyle is changed into blood.

Sclerotica. The name of one of the coats of the eye.

Scurf. Small exfoliations of the cuticle, which take place after some eruptions on the skin, a new cuticle being formed underneath during the exfoliation.

Secretion. This word is used to express the separation of various fluids from the blood, while the latter either preserves its chemical properties, or disperses after its elements have undergone another order of combinations.

Sentient. This term is applied to those parts which are more susceptible of feeling than others, as the sentient extremities of the nerves, &c.

Soporiferous, or *Soporific.* A word used to denote whatever is capable of causing sleep.

Specific. A remedy which, for a certain train of symptoms, is, in all cases and under all circumstances, curative.

Splint. A long piece of wood, tin, or strong pasteboard, used to prevent the ends of broken bones from moving so as to disturb the process of adhesion.

Sporadic. A word applied to such infections and other diseases as seize a few persons at any time or season.

Stimulus. That which possesses the power of exciting vital energy.

Stupor. Insensibility.

Strumous. Of the nature of scrofula.

Subcutaneous. Under the true skin.

Sublimation. A process by which volatile substances are raised by heat, and again condensed in a solid form, — differing from evaporation only in being confined to solid substances.

Subsultus Tendinum. Weak, convulsive motions or twitchings of the tendons, mostly of the hands, generally observed in the extreme stages of putrid fever.

Sudorific. That which causes perspiration.

Suppuration. That diseased action, succeeding inflammation, by which pus is deposited in tumors.

Syncope. Fainting or swooning.

Synovia. An unctuous fluid secreted from certain glands within the joints, serving the purpose of lubricating the cartilaginous surfaces of the articulations, thereby facilitating the movement of the joints.

Therapeutics. That branch of medicine which treats of the different means employed for curing diseases, and of the application of those means.

Torpor. A numbness or deficient sensation.

Traumatic. When applied to diseases, signifies those which are the result of wounds.

Trituration. The act of reducing in a mortar solid substances to subtile powder. By this separation of molecules, it is believed that the active medicinal principle is thoroughly developed.

Tubercles. Small, round, hard elevations, which are found in diseased lungs and other organs. The term is also applied to certain cutaneous eruptions.

Tympanitic. Applied to an elastic, distended state of the abdomen, sounding like a drum when struck, and which is caused by an accumulation of air.

Typhoid. A term expressing a species of low fever, characterized by debility.

Varicose. Applied to veins unnaturally distended with blood.

Vertigo. Giddiness.

Vesicle. An elevation of the external skin or cuticle, containing a transparent fluid.

Ventricle. A term given by anatomists to the cavities of the brain and heart.

30 *

THE DISEASES

?OR WHICH EACH REMEDY IS ADMINISTERED,
WITH ITS ANTIDOTE.

---

1. ACONITE. — *Antidotes,* **Camphor, Nux Vomica.**

Page 45, Teething; 50, Tic Douloureux; 68, Dysente y; 84, Catarrh; 92, Laryngitis; 93, Bronchitis; 95, Pneumonia; 97, Pleurisy; 99, Hæmoptysis; 100, Consumption; 103, Carditis; 104, Angina Pectoris; 109, Headache; 115, Vertigo; 116, Phrenitis; 117, Apoplexy; 119, Hydrocephalus; 120, Epilepsy; 126, Neuralgia; 134, Simple Fever; 135, Inflammatory Fever; 136, Typhus Fever; 142, Small Pox; 147, Chicken Pox; 148, Measles; 150, Scarlet Fever; 152, Scarlet Rash; 153, Miliary Fever; 155, Erysipelas; 165, Crusta Lactea; 169, Whitlow; 176, Ophthalmia; 177, Sty; 184, Otitis; 192, Gout; 201, Epistaxis; 202, Nephritis; 203, Cystitis; 258, Fainting.

2. ANTIMONIUM TARTARIZATUM. — *Antidote,* **Hepar Sulphuris.**

138, Bilious Fever; 139, Intermittent Fever; 142, Small Pox.

3. ARNICA. — *Antidotes,* **Camphor, Ignatia.**

59, Gastritis; 97, Pleurisy; 99, Hæmoptysis; 122, Paralysis; 130, Tetanus; 167, Bile; 192, Gout; 202,

Nephritis ; 228, Wounds ; 238, Dislocations ; 243, Fractures ; 251, Asphyxia.

4. ARSENICUM. — *Antidotes*, Camphor, Ipecacuanha.

50, Tic Douloureux ; 51, Aphthæ ; 55, Ulcerated Sore Throat ; 59, Gastritis ; 61, Hæmatemesis ; 70, Cholera ; 71, Asiatic Cholera ; 78, Splenitis ; 78, Enteritis ; 85, Influenza ; 98, Asthma ; 103, Carditis ; 123, Convulsions ; 139, Intermittent Fever ; 142, Small Pox ; 155, Erysipelas ; 172, Ulcers ; 193, Scurvy ; 195, Dropsy.

5. BELLADONNA. — *Antidotes*, Coffea, Hepar Sulphuris, Pulsatilla.

45, Teething ; 50, Tic Douloureux ; 52, Mumps ; 53, Inflammatory Sore Throat ; 55, Ulcerated Sore Throat ; 76, Hepatitis ; 84, Catarrh ; 85, Influenza ; 86, Cough ; 88, Whooping Cough ; 93, Bronchitis ; 95, Pneumonia ; 97, Pleurisy ; 111, Congestive Headache ; 115, Vertigo ; 116, Brain Fever ; 117, Apoplexy ; 120, Epilepsy ; 123, Convulsions; 126, Neuralgia; 135, Inflammatory Fever; 150, Scarlet Fever ; 153, Miliary Fever ; 176, Ophthalmia ; 181, Amaurosis ; 185, Otalgia ; 201, Epistaxis.

6. BROMINE.

90, Croup.

7. BRYONIA. — *Antidotes*, Aconite, Ignatia, Chamomilla.

56, Dyspepsia ; 60, Cardialgia ; 64, Constipation ; 68, Dysentery ; 76, Hepatitis ; 86, Cough ; 93, Bronchitis ; 95, Pneumonia ; 97, Pleurisy ; 119, Hydrocephalus ; 122, Paralysis ; 136, Typhus Fever ; 145, Variola ; 148, Measles ; 152, Scarlet Rash ; 189, Rheumatism ; 221, Chlorosis.

8. CALCAREA CARBONICA. — *Antidotes*, Aconite, Chamomilla.

45, Teething; 100, Consumption; 111, Congestive Headache; 114, Chronic Headache; 120, Epilepsy; 169, Chilblains; 171, Warts; 177, Sty; 199, Scrofula; 220, Leucorrhœa.

9. CAMPHOR. — *Antidote*, Opium.

71, Asiatic Cholera.

10. CANTHARIS. — *Antidote*, Camphor.

202, Nephritis; 203, Cystitis; 204, Urinary Difficulties.

11. CARBO VEGETABILIS. — *Antidotes*, Arsenicum, Coffea.

136, Typhus Fever; 138, Bilious or Gastric Fever; 138, Yellow Fever; 180, Near-Sightedness; 193, Scurvy.

12. CHAMOMILLA. — *Antidotes*, Aconite, Pulsatilla.

45, Teething; 48, Toothache; 56, Dyspepsia; 63, Heartburn; 66, Diarrhœa; 70, Cholera; 75, Jaundice; 84, Catarrh; 112, Rheumatic Headache; 113, Nervous Headache; 123, Convulsions; 138, Bilious Fever; 185, Earache; 211, Colic; 212, Sleeplessness; 215, Infantile Diarrhœa; 216, Cholera Infantum; 220, Menorrhagia; 258, Fainting.

13. CINCHONA. — *Antidotes*, Arnica, Belladonna, Sulphur.

75, Jaundice; 78, Splenitis; 99, Hæmoptysis; 139, Intermittent Fever; 186, Deafness; 195, Dropsy; 258, Fainting.

14. COCCULUS. — *Antidotes*, Camphor, Nux Vomica.

198, Sea Sickness.

15. COFFEA. — *Antidotes*, Aconite, Chamomilla.

45, Teething; 113, Nervous Headache; 125, Chorea; 126, Neuralgia; 128, Hysteria; 147, Chicken Pox; 152, Scarlet Rash; 153, Miliary Fever; 212, Sleeplessness.

16. COLOCYNTH. — *Antidote*, Camphor.

50, Tic Douloureux; 62, Colic; 68, Dysentery.

17. CUPRUM. — *Antidotes*, Belladonna, Nux Vomica.

71, Asiatic Cholera; 88, Whooping Cough.

18. DROSERA. — *Antidote*, Camphor.

88, Whooping Cough; 180, Far-Sightedness.

19. DULCAMARA. — *Antidotes*, Camphor, Ipecacuanha.

66, Diarrhœa; 186, Deafness; 195, Dropsy.

20. HELLEBORUS. — *Antidotes*, Camphor, Cinchona.

119, Hydrocephalus; 181, Amaurosis; 195, Dropsy.

21. HEPAR SULPHURIS. — *Antidotes*, Belladonna, Chamomilla.

89, Croup; 92, Laryngitis; 92, Hoarseness; 171, Abscess.

22. HYOSCYAMUS. — *Antidotes*, Cinchona, Belladonna.

86, Cough; 116, Phrenitis; 128, Hysteria; 178, Strabismus.

23. IGNATIA. — *Antidotes*, Pulsatilla, Chamomilla.

45, Dentition; 53, Quinsy; 123, Convulsions; 128, Hysteria; 258, Fainting.

24. IPECACUANHA. — *Antidotes*, Arsenicum, Arnica.

61, Hæmatemesis; 68, Dysentery; 70, Cholera; 71,

Asiatic Cholera; 86, Cough; 98, Asthma; 110, Headache from Indigestion; 189, Fever and Ague; 148,
Measles; 198, Seasickness; 214, Vomiting; 216, Cholera Infantum; 220, Menorrhagia.

25. LACHESIS.—*Antidotes*, Nux Vomica, Belladonna.

53, Quinsy; 89, Croup; 92, Laryngitis; 92, Hoarseness; 168, Anthrax.

26. LYCOPODIUM.—*Antidotes*, Camphor, Pulsatilla.

50, Tic Douloureux; 64, Constipation; 100, Con
sumption; 167, Psoriasis.

27. MERCURIUS SOL.—*Antidotes*, Belladonna, Sulphur.

48, Toothache; 51, Aphthæ; 52, Mumps; 53, Inflammatory Sore Throat; 55, Ulcerated Sore Throat;
68, Dysentery; 75, Jaundice; 79, Worms; 85, Influenza; 142, Variola.

28. MERCURIUS COR.

69, Dysentery.

29. NUX VOMICA.—*Antidotes*, Camphor, Pulsatilla.

48, Toothache; 56, Dyspepsia; 60, Cardialgia; 62,
Colic; 64, Constipation; 73, Hemorrhoids; 75, Jaun-
dice; 84, Catarrh; 86, Cough; 109, Headache; 129,
Delirium Tremens; 138, Bilious Fever; 192, Gout;
181, Amaurosis; 211, Infantile Colic; 213, Infantile
Catarrh; 220, Menorrhagia; 258, Fainting.

30. OPIUM.—*Antidotes*, Camphor, Sulphur.

64, Constipation; 117, Apoplexy; 129, Delirium
Tremens; 258, Fainting.

31. PHOSPHORUS. — *Antidotes*, **Camphor, Coffea.**

76, Hepatitis; 95, Pneumonia; 100, Consumption; 186, Deafness.

32. PULSATILLA. — *Antidotes*, **Chamomilla, Coffea.**

48, Toothache; 56, Dyspepsia; 59, Gastritis; 60, Cardialgia; 64, Constipation; 66, Diarrhœa; 84, Catarrh; 86, Cough; 91, Millar's Asthma; 99, Hæmoptosis; 110, Headache from Indigestion; 113, Nervous Headache; 113, Sick Headache; 128, Hysteria; 148, Measles; 177, Sty; 180, Weakness of Sight; 185, Earache; 186, Deafness; 189, Rheumatism; 192, Gout; 218, Amenorrhœa; 220, Leucorrhœa; 221, Chlorosis.

33. RHUS TOXICODENDRON. — *Antidotes*, **Belladonna, Sulphur.**

136, Typhus Fever; 154, Nettle Rash; 155, Erysipelas; 160, Ringworm; 163, Lichen; 164, Acne; 189, Rheumatism; 214, Tooth Rash.

34. SAMBUCUS. — *Antidotes*, **Arsenicum, Camphor.**

91, Asthma of Millar; 213, Catarrh.

35. SANTONINE.

80, Worms.

36. SEPIA. — *Antidote*, **Aconite.**

114, Chronic Headache; 160, Ringworm; 162, Herpes; 163, Lichen; 218, Amenorrhœa; 220, Leucorrhœa.

37. SILICEA. — *Antidote*, **Camphor.**

163, Porrigo; 164, Acne; 168, Carbuncle; 172, Ulcers.

38. SPONGIA. — *Antidote*, **Camphor.**

89, Croup; 92, Laryngitis.

39. STANNUM. — *Antidote*, Pulsatilla.

79, Worms ; 93, Bronchitis ; 100, Consumption.

40. STRAMONIUM. — *Antidotes*, Belladonna, Nux Vomica.

116, Inflammation of the Brain.

41. SULPHUR. — *Antidotes*, Aconite, Pulsatilla.

51, Aphthæ ; 66, Diarrhœa ; 73, Hemorrhoids ; 79,
Worms ; 97, Pleuritis ; 100, Consumption ; 114, Chronic
Headache ; 119, Hydrocephalus ; 122, Paralysis ; 142,
Variola ; 162, Herpes ; 169, Chilblains ; 171, Warts ;
172, Ulcers ; 193, Scurvy ; 199, Scrofula ; 214, Gum
Rash ; 215, Infantile Diarrhœa ; 218, Amenorrhœa ;
220, Leucorrhœa ; 221, Chlorosis.

42. SULPHURIC ACID. — *Antidote*, Pulsatilla.

180, Weakness of Sight ; 193, Scurvy.

43. VERATRUM. — *Antidotes*, Aconite, Coffea.

70, Cholera ; 71, Asiatic Cholera ; 78, Enteritis ; 88
Whooping Cough ; 113, Sick Headache ; 126, Neuralgia ; 216, Cholera Infantum.

NOTE. The antidotes above mentioned are to be resorted to when an
aggravation of the disease is obviously caused by the action of the remedy
used. Such medicinal aggravation is to be recognized by a sudden,
instead of a gradual increase of feverish or other symptoms, and also
by indications characteristic of the medicine, that had not been previously observed. It is, however, seldom if ever necessary to resort
to antidotes, especially under the administration of the globules herein
advised, excepting in very rare cases of an excessive susceptibility to
medicinal influence.

APPENDIX.

REMEDIES FOR EXTERNAL APPLICATION.

WE append to this edition a list of the commonly accepted tinctures recognized as valuable for external use. There is no reason to doubt that a beneficial influence is exercised on certain diseases by the application of a remedy to the extensively absorbing surface of the body in conjunction with its internal use. The following are kept in the form of tincture, to be diluted as directed below : —

ARNICA.
(*Montana.*)

The best preparation of this medicine for outward application is a proportion of one part of the tincture to six parts of water. It is to be applied to *bruises* and wounds of all kinds, where the skin remains unbroken. It is also made use of to check the *boils* and *carbuncles* in the inflammatory or first stage, as well as in *fractures* and *dislocations*, after the bones have been replaced.

31   (361)

CALENDULA.
(*Officinalis.*)

This is prepared for external application by mixing the tincture with water. in the proportion of one part of the tincture to eight parts of water. It is used for *wounds* made by a sharp surface, by which the skin is cut or lacerated.

RHUS.
(*Toxicodendron.*)

This is prepared, as above, by mixing the tincture with water. The proportion of the tincture to the water should be as one to ten; for example, a tea-spoonful of the tincture of rhus to five table-spoonfuls of water. It is beneficial in rheumatism, applied to the parts affected with pain; also in *sprains* and *injuries to joints.*

ACONITE.

This tincture, in the proportion of three or four drops to a table-spoonful of warm water, may be applied with very good effect in *ophthalmia,* or *inflammation of the eyes,* particularly if the local inflammation is attended with general febrile symptoms. A few drops of the tincture mixed with equal parts of water, if rubbed over the seat of pain in *neuralgia* or *tic douloureux,* will frequently afford great relief, in connection with the internal use of three or four globules of aconite, taken every fifteen minutes.

BELLADONNA.

A tea-spoonful of this tincture put into half a pint of warm water, has been found of much service in *con-*

gestive headaches and in *inflammation of the breasts.*
A linen cloth wet with the solution is to be laid upon
the inflamed surface. When used in this manner,
three or four globules of the same medicine may be
taken internally. The remarkable action of belladonna on the iris is well known.

For greater convenience, we add here, in tabular
form, the affections for which the above named medicines are used, externally, with the strength of the
lotions : —

Affections.	*Remedy.*	*Strength.*
BEDSORES, ...	Arnica, . . .	1 part of tincture to 10 parts of water.
BITES OF INSECTS, .	Arnica, . . .	1 " " to 10 " "
BOILS, ...	Arnica, . . .	1 " " to 10 " "
BREASTS, Inflam. of, .	Belladonna, .	1 " " to 20 " "
BUNIONS, ...	Arnica, . . .	1 " " to 10 " "
BURNS, ...	Arnica, . . .	1 " " to 10 " "
CHILBLAINS, ...	Arnica, . . .	1 " " to 6 " "
CONTUSIONS, ...	Arnica, . . .	1 " " to 6 " "
CORNS, ...	Arnica, . . .	1 " " to 4 " "
CUTS, ...	Calendula, .	1 " " to 8 " "
CARBUNCLES, ...	Arnica, . . .	1 " " to 6 " "
DISLOCATIONS.	Arnica, . . .	1 " " to 8 " "
EYES, Blackened, . . .	Arnica, . . .	1 " " to 8 " "
" Inflamed, ...	Aconite, . .	1 " " to 12 " "
FRACTURES, ...	Arnica, . . .	1 " " to 8 " "
NEURALGIA, ...	Aconite, . .	1 " " to 6 " "
RHEUMATISM, ...	Rhus T. .	1 " " to 4 " "
SPRAINS, ...	Rhus T. . .	1 " " to 4 " "
WOUNDS, Incised, . . .	Calendula, .	1 " ' to 12 " "
" Contused, . .	Arnica, . . .	1 " " to 8 ' "

DIPHTHERIA.

THIS is a disease of the throat, or rather a constitu‑
tional affection of an infectious nature, that manifests
itself especially by white, or yellowish white spots on
the inside of the throat, and on the tonsils, somewhat
similar to the membrane of croup, excepting that the
latter is a fibrinous instead of an albuminous exuda‑
tion, and begins to form lower down the passage
leading to the lungs. The presence of this membrane
in diphtheria is attended with difficulty of swallowing,
a sensation as of a hard substance in the throat, a swol‑
len state of the glands, like the mumps, an extremely
foul breath, a rapid pulse, and great debility. As the
disease progresses, the breathing becomes more labo‑
rious and rattling; a hollow and whistling cough is
heard, like the metallic sounding cough of croup; vio‑
lent pains are often felt in the limbs; the pulse increases
in rapidity, and suffocation seems imminent.

In the first stages of this disease, when the usual
symptoms are present, and the yellowish white mem
brane is observed in the throat, the medicine indi
cated is

ACONITE.

One drop of the tincture in a tea-spoonful of water
given every fifteen minutes. (364)

After a few doses of the above, administered to lessen the general febrile action, and if the membranous exu dation appears to be extending, the

TINCTURE OF IODINE

should be inhaled from an open vessel placed near the patient, and it should be also used, diluted with equal parts of water, as a gargle. In an advanced stage, where there is great debility, and a puffy swelling of the throat, give

ARSENICUM.

Bromine, bichromate of potass, caustic ammonia, muriate of iron, and other remedies of the kind, are recommended, in strong doses, with a view to the detachment and removal of the membrane which threatens suffocation, as in membranous croup.

In consequence of the grave character of Diphtheria, the serious mechanical impediment to respiration, which an extension of the membrane into the larynx would present, the great importance of a prompt administration of the suitable medicine, as well as the unsettled condition of remedial measures in relation to a disease of comparatively recent origin, or, at least, development in this country, it is not, except in its most simple form, a suitable subject for domestic management, and no inexperienced person should venture upon its treatment.

DIET. — This should consist of strong beef tea, a tea-spoonful every hour, after the febrile condition is over, as great prostration of strength is peculiar to diphtheria, and sustenance from nourishing food is of the highest importance. Wine whey may be given often to children, and brandy and water in small quantities, but frequently repeated, to adults.

PROPHYLACTICS.

(Preventives of Disease.)

BY B. L. HILL, M. D., SURGEON IN U. S. ARMY.

TO PREVENT SCARLET FEVER.

GIVE Belladonna, at the third attenuation, three to six pellets, according to the age of the child, every morning, during the prevalence of the epidemic. This is for the common or mild form of the disease. If the prevailing epidemic is of the *malignant* kind, producing fatal ulcerations of the throat, give *belladonna* once in two days, and *mercurius corrosivus* at the third attenuation on the alternate day.

While *belladonna* is a very certain preventive of the common eruptive scarlatina, it is not as certain to prevent the *malignant* form. Though it renders the latter much more mild, the *mercurius corrosivus* is necessary to ward it off entirely, or so modify as to divest it of the dangerous features.

TO PREVENT YELLOW FEVER.

Take *aconite, belladonna*, and *macrotin*, first in rotation, one dose a day. If there is any headache, or pains occur in other parts of the body, or a languid feeling, take a dose twice ᴏr three times a day in rotation.

TO PREVENT BILIOUS FEVER OR AGUE.

Take *podophyllin*, *baptisia*, and *gelseminum*, first in rotation, one dose at night, and if symptoms of fever, as headache and loss of appetite, or bad taste in the mouth in ·the morning appear, take a dose three times a day, and refrain entirely from food for one or two days.

TO PREVENT TYPHOID FEVER.

When exposed, as in nursing the sick, take *baptisia* second, and *macrotin* second, a dose three times a day.

TO PREVENT SMALL POX.

Use *macrotin* first, night and morning, and if nursing, or exposed frequently, use it every four hours.

TO PREVENT CHOLERA.

Camphor (pellets medicated with the pure tincture). *Veratrum* third, and *arsenicum* third, should be taken in rotation ; a dose morning, noon and night, in the order named, so as to take a dose of each every twenty-four hours. If any sense of weakness or trembling comes on, use the *camphor* oftener; if pain or uneasiness in the bowels, threatening diarrhœa, use the *veratrum*, and for increased thirst, with uneasiness at the stomach, *arsenicum* more frequently.

TO PREVENT DIARRHŒA.

Where it is prevailing as an epidemic, *ipecacuanha*

at night, and *veratrum* in the morning, will often suffice. For teething children, give *ipecacuanha* and *chamomilla* in the same manner.

TO PREVENT DYSENTERY.

In hot weather, when bilious diseases prevail, use *mercurius* third, *podophyllin* second, and *leptandrin* first, in rotation, giving one dose a day.

In the winter, or when typhoid fevers prevail, use *mercurius* and *rhus toxicodendron* alternately, a dose every day.

TO PREVENT ITCH.

A dose of *sulphur*, or rubbing a little *flour of sulphur* on the hands, will generally suffice.

TO PREVENT COLDS.

Keep the arms, hands, and chest well clothed and warm. Affecting the head as catarrh, or the pelvic regions, keep the feet and ankles warm and dry. Affecting joints and muscles, as rheumatism, protect the spine (back) from colds and currents of air.

After an accidental exposure, as by getting the feet wet, or being caught in a shower, drink bountifully of cold water, and take a dose of *nux*, followed in an hour by *aconite*, if any chilliness is felt, or *copaiva*, if the head is "stuffed up."

In winter and spring, when the weather is mild, but there is snow, or the ground is damp, more clothes are necessary than when it is freezing hard and the air is dry.

INDEX.